MOTHER OPOSSUM AND HER BABIES

Teacher's Guide
Preschool–1

Skills
Observing, Comparing, Describing, Communicating, Role Playing, Estimating, Counting, Measuring, Ordering, Problem Solving, Recording

Concepts
Animal Behaviors (feeding, defense, parenting), Animal Growth and Development, Life Cycle, Marsupials, Form and Function, Estimation, Number Sense, Place Value, Measurement

Themes
Patterns of Change, Structure, Diversity and Unity, Scale

Mathematics Strands
Logic and Language, Number, Measurement, Statistics

Nature of Science and Mathematics
Real-Life Applications, Interdisciplinary

by

Jean C. Echols, Jaine Kopp, Ellen Blinderman, Kimi Hosoume

LHS GEMS

GEMS
Great Explorations in Math and Science
Lawrence Hall of Science
University of California at Berkeley

Lawrence Hall of Science
    Director: Ian Carmichael

Original funding for GEMS was provided by the A.W. Mellon Foundation and the Carnegie Corporation of New York, with equipment donations from Apple Computer. Under a grant from the National Science Foundation, GEMS Leader's Workshops were held throughout the United States. GEMS has received grants from the people at Chevron USA; the Hewlett Packard Company; the McDonnell-Douglas Employee's Community Fund and McDonnell-Douglas Foundation; Employees Community Fund of Boeing California and the Boeing Corporation; Join Hands, the Health and Safety Educational Alliance; the Microscopy Society of America (MSA); the Shell Oil Company Foundation; the Crail-Johnson Foundation; and the William K. Holt Foundation. This support does not imply responsibility for statements or views expressed in publications of the GEMS program.

Under a grant from the National Science Foundation, GEMS Leader's Workshops have been held across the country. For further information on GEMS leadership opportunities, or to receive a publication brochure and the *GEMS Network News*, please contact GEMS at the address and phone number below.

Development of this guide was sponsored in part by the Department of Education Fund for the Improvement of Post-Secondary Education (FIPSE) and a grant from the National Science Foundation.

COMMENTS WELCOME

GEMS guides are revised periodically to incorporate teacher comments and new approaches. We welcome your criticisms, suggestions, helpful hints, and any anecdotes about your experience presenting GEMS activities. Your suggestions are reviewed each time a GEMS guide is revised. Please send your comments to:

University of California, Berkeley
GEMS Revisions
Lawrence Hall of Science # 5200
Berkeley, CA 94720-5200

Phone: (510) 642-7771
Fax: (510) 643-0309
Email: GEMS@uclink4.berkeley.edu
Website: www.lhs.berkeley.edu/GEMS

GEMS STAFF

Director
    Jacqueline Barber
Associate Director
    Kimi Hosoume
Associate Director/Principal Editor
    Lincoln Bergman
GEMS Network Director
    Carolyn Willard
GEMS Workshop Coordinator
    Laura Tucker
Mathematics Curriculum Specialist
    Jaine Kopp
Development Specialists
    Lynn Barakos, Katharine Barrett,
    Kevin Beals, Ellen Blinderman,
    Beatrice Boffen, Gigi Dornfest,
    John Erickson, Stan Fukunaga,
    Philip Gonsalves, Linda Lipner,
    Karen Ostlund, Debra Sutter
Workshop Administrator
    Terry Cort
Materials Manager
    Vivian Tong
Financial Assistant
    Alice Olivier
Distribution Coordinator
    Karen Milligan
Distribution Representative
    Felicia Roston
Shipping Assistant
    Jodi Harskamp
Director of Marketing and Promotion
    Matthew Osborn
Senior Editor
    Carl Babcock
Editor
    Florence Stone
Public Information Representative
    Gerri Ginsburg
Principal Publications Coordinator
    Kay Fairwell
Art Director
    Lisa Haderlie Baker
Senior Artists
    Carol Bevilacqua, Lisa Klofkorn, Rose Craig
Staff Assistants
    Larry Gates, Trina Huynh,
    Chastity Pérez, Christine Tong,
    Dorian Traube

Great Explorations in Math and Science (GEMS) Program

GEMS, Great Explorations in Math and Science, is an ongoing curriculum development project at the Lawrence Hall of Science—a public science center of the University of California at Berkeley.

LHS offers a full program of activities for the public, including workshops and classes, exhibits, films, lectures, and special events. LHS is also a center for teacher education and curriculum research and development. Over the years, LHS staff has developed a multitude of activities, assembly programs, classes, and interactive exhibits. These programs have proven successful at LHS and should be useful to schools, other science centers, museums, and community groups. A number of these guided-discovery activities are published as a GEMS title after an extensive refinement process that includes classroom testing, ensuring the use of easy-to-obtain materials, and carefully written step-by-step instructions and background information to allow presentation by teachers without special background in mathematics or science.

Contributing Authors

Jacqueline Barber
Katharine Barrett
Kevin Beals
Lincoln Bergman
Ellen Blinderman
Beverly Braxton
Kevin Cuff
Linda De Lucchi
Gigi Dornfest
Jean Echols
John Erickson
Philip Gonsalves
Jan M. Goodman
Alan Gould
Catherine Halversen
Kimi Hosoume
Sue Jagoda
Jaine Kopp
Linda Lipner
Larry Malone
Cary Sneider
Craig Strang
Debra Sutter
Jennifer Meux White
Carolyn Willard

Title Typeface
The typeface used for the title on the cover and title page is Renoir—a facsimile of the handwriting of the French impressionist painter Auguste Renoir.

ACKNOWLEDGMENTS

Photographs: **Richard Hoyt**
Cover: **Lisa Haderlie Baker**
Illustrations: **Lisa Haderlie Baker, Rose Craig**

Thanks to LHS staff members **Katharine Barrett** and **Beatrice Boffen** for their insightful contributions to *Mother Opossum and Her Babies*.

Thanks to UC Berkeley Professor **Reginald H. Barrett** of the Environmental Science department for his scientific review of this guide, and to **Florence Stone** for her assistance in researching opossums.

Thanks to the National Science Foundation for funding the guide's early development through its Teacher Enhancement Program.

Thanks to **Linette Mace**, and her first grade class, for the opossum poem on page 1. Thanks also go to **Mary Ehler** and to **Ellis Elliott** for their individual song lyrics on page 61.

Special thanks to teachers **Jeanne Wang** and **Marey Donnelly** of Emerson Elementary School in Berkeley, California, for allowing us to photograph their classes. And, an extra special thank you to the Emerson kindergarteners who enliven the photos in this guide. Thanks also to River Legacy Nature School Director **Susie Barcus** for allowing us to use photos of her Arlington, Texas, students engaged in opossum activities.

The authors extend their deep appreciation to the many teachers who presented these lessons to hundreds of children across the country. It is clear by your enthusiastic letters that you all contributed to a better understanding and respect for this unique animal. We hope you continue to enjoy opossums!

Reviewers

Many thanks to the following educators who reviewed, tested, or coordinated the reviewing of this series of GEMS/PEACHES materials in manuscript and draft form. Their critical comments and recommendations, based on presentation of these activities nationwide, contributed significantly to these GEMS publications. Their participation in the review process does not necessarily imply endorsement of the GEMS program or responsibility for statements or views expressed. Their role is an invaluable one, and their feedback is carefully recorded and integrated as appropriate into the publications.

ALASKA

Pacific Northern Academy,
Anchorage
Naomi Mayer
(Coordinator)
Laura Schue
Kristin Hewitt
Fay Sims

CALIFORNIA

4C's Children's Center, Oakland
Yolanda
Coleman-Wilson

24 Hour Children Center,
Oakland
Sheryl Lambert
Ella Tassin
Inez Watson

Afterschool Program, Piedmont
Willy Chen

Alameda Head Start, Alameda
Michelle Garabedian
Debbie Garcia
Stephanie Josey

Albany Children's Center,
Albany
Celestine Whittaker

Bancroft School, Berkeley
Cecilia Saffarian

Bartell Childcare and Learning
Center, Oakland
Beverly Barrow
Barbara Terrell

Belle Vista Child Development
Center, Oakland
Satinder Jit K. Rana

Berkeley-Albany YMCA,
Berkeley
Trinidad Caselis

Berkeley Hills Nursery School,
Berkeley
Elizabeth Fulton

Berkeley/Richmond Jewish
Community, Berkeley
Terry Amgott-Kwan

Berkwood-Hedge School,
Berkeley
Elizabeth Wilson

Bernice & Joe Play School,
Oakland
Bernice
Huisman-Humbert

Bing School, Stanford
Kate Ashbey

Brookfield Elementary School,
Oakland
Kathy Hagerty

Brookfield Head Start, Oakland
Suzie Ashley

Butte Kiddie Corral, Shingletown
Cindy Stinar Black

Cedar Creek Montessori,
Berkeley
Idalina Cruz
Jeanne Devin
Len Paterson

Centro Vida, Berkeley
Rosalia Wilkins

Chinese Community United
Methodist Church, Oakland
Stella Ko Kwok

Clayton Valley Parent Preschool,
Concord
Lee Ann Sanders
Patsy Sherman

Compañeros del Barrio State
Preschool, San Francisco
Anastasia Decaristos
Laura Todd

Contra Costa College, San Pablo
Sylvia Alvarez-Mazzi

Creative Learning Center,
Danville
Brooke H. B. D'Arezzo

Creative Play Center,
Pleasant Hill
Debbie Coyle
Sharon Keane

Dena's Day Care, Oakland
Kawsar Elshinawy

Dover Preschool, Richmond
Alice J. Romero

Duck's Nest Preschool, Berkeley
Pierrette Allison
Patricia Foster
Mara Ellen Guckian
Ruth Major

East Bay Community Children's
Center, Oakland
Charlotte Johnson
Oletha R. Wade

Ecole Bilingue, Berkeley
Nichelle R. Kitt
Richard Mermis
Martha Ann Reed

Emerson Elementary, Berkeley
Jeanne Wang
(Coordinator)
Marey Donnelly
Kim Etzel

Emerson Elementary School,
Oakland
Pamela Curtis-Horton

Emeryville Child Development
Center, Emeryville
Ellastine Blalock
Jonetta Bradford
William Greene
Ortencia A. Hoopii

Enrichment Plus Albert Chabot
School, Oakland
Lisa Dobbs

Family Day Care, Oakland
Cheryl Birden
Penelope Brody
Eufemia Buena Byrd
Mary Waddington

Family Day Care, Orinda
Lucy Inouye

Gan Hillel Nursery School,
Richmond
Denise Moyes-Schnur

Gan Shalom Preschool, Berkeley
Iris Greenbaum

Garner Toddler Center, Alameda
Uma Srinath

Gay Austin, Albany
Sallie Hanna-Rhyne

Giggles Family Day Care,
Oakland
Doris Wührmann

Greater Richmond
Social Services Corp., Richmond
Lucy Coleman

Happy Lion School, Pinole
Sharon Espinoza
Marilyn Klemm

Hintil Kuu Ca Child
Development Center, Oakland
Eunice C. Blago
Kathy Moran
Gina Silber
Agnes Tso
Ed Willie

Jack-in-the-Box Junction
Preschool, Richmond
Virginia Guadarrama

Kinder Care, Oakland
Terry Saugstad

King Preschool, Richmond
Charlie M. Allums

The Lake School, Oakland
Margaret Engel
Patricia House
Vickie Stoller

Learning Adventures
Child Development, Redding
Dena Keown

Longfellow Child Development
Center, Oakland
Katryna Ray

Los Medanos Community
College, Pittsburg
Judy Henry
Filomena Macedo

Maraya's Developmental Center,
Oakland
Maria A. Johnson-Price
Gayla Lucero

Mark Twain School Migrant
Education, Modesto
Grace Avila

Mary Jane's Preschool,
Pleasant Hill
Theresa Borges

Merritt College Children's
Center, Oakland
Deborah Green
Virginia Shelton

Mickelson's Child Care, Ramona
Levata Mickelson

Mills College Children's Center,
Oakland
Monica Grycz

Mission Head Start,
San Francisco
Pilar Marroquin
Mirna Torres

The Model School
Comprehensive, Berkeley
Jenny Schwartz-Groody

Montclair Community
Play Center, Oakland
Elaine Guttmann
Nancy Kliszewski
Mary Loeser

Next Best Thing, Oakland
Denise Hingle
Franny Minervini-Zick

Oak Center Christian Academy,
Oakland
Debra Booze

Oakland Parent Child Center,
Oakland
Barbara Jean Jackson

Orinda Preschool, Orinda
Tracy
Johansing-Spittler

Oxford St. Learning Road,
Berkeley
Vanna Maria Kalofonos

Peixoto Children's Center,
Hayward
Alma Arias
Irma Guzman
Paula Lawrence
Tyra Toney

Piedmont Cooperative
Playschool, Piedmont
Marcia Nybakken

Playmates Daycare, Berkeley
Mary T. McCormick

Rainbow School, Oakland
Mary McCon
Rita Neely

Redwood Heights Elementary,
Oakland
Linda Rogers

San Antonio Head Start,
Oakland
Cynthia Hammock
Ilda Terrazas

Sequoia Nursery School,
Oakland
Karen Fong

Sequoyah Community Preschool,
Oakland
Erin Smith
Kim Wilcox

Shakelford Head Start, Modesto
Teresa Avila

St. Vincent's Day Home,
Oakland
Pamela Meredith

Sunshine Preschool, Berkeley
Poppy Richie

U. C. Berkeley Child Care
Services Smyth Fernwald II,
Berkeley
Diane Wallace
Caroline W. Yee

Walnut Ave. Community
Preschool, Walnut Creek
Evelyn DeLanis

Washington Child Development
Center, Berkeley
Heather Jones

Washington Kids Club, Berkeley
Adwoa A. Mante

Westview Children's Center,
Pacifica
Adrienne J. Schneider

Woodroe Woods, Hayward
Wendy Justice

Woodstock Child Development
Center, Alameda
Mary Raabe
Denise M. Ratto

Woodstock School, Alameda
Amber D. Cupples

Yuk Yan Annex, Oakland
Eileen Lok

YWCA Oakland, Oakland
Iris Ezeb
Grace Perry

IOWA

Area Education Agency 6 ESD,
Marshalltown
Linda Carter McCartney
(Coordinator)

Pineview Elementary, Iowa Falls
Shelly Klaver
Mary Lou Topp

Hubbard-Radcliffe Elementary,
Radcliffe
Linette Mace

MISSOURI

Franklin Elementary,
Cape Girardeau
Winona Crampton
(Coordinator)
Tammy Raddle
Barbara Egbert
Cheryl Hendershott

NEVADA

Sutro Elementary, Dayton
Michele Paul
(Coordinator)
Nancy Scott
Cindy Wilde
Jill Howe

TEXAS

Armand Bayou Elementary
School, Houston
Myra Luciano

Central Elementary, Henderson
Pam Glenn

Henderson ISD, Henderson
Gina Gardiner
(Coordinator)

Henderson Kindergarten,
Henderson
Sara Hood
(Coordinator)
Chrystal Floyd
Pam Smith
Kerri Miller

Living Science Center, Arlington
Susan Barcus
(Coordinator)
Terry Carver
Linda Watts
Patty Filewood
Debbie Vernon

River Legacy Nature School,
Arlington
Becky Nussbaum
Ellis Elliott

CONTENTS

Introduction

Many interesting and appealing attributes and behaviors make the opossum a great animal to learn about with young children. Although some people mistakenly think of opossums as big rats, they differ in many ways from rats and other rodents. Opossums belong to the order *marsupialia* (pouched mammals), which includes kangaroos and koalas. Opossums have the distinction of being the only marsupial found in the wild in North America. They are widely found in the United States, and because they are well adapted to survive in all types of habitats—from forests and farmland to suburbs and cities—their range continues to expand.

Opossums are mainly active at night, yet children have seen them—perhaps from the car on a dark road or in their yards as the animals rummage in the trash or eat from the cat's bowl. Often, community zoos and nature centers have opossums to observe and handle. Some children may have seen only dead opossums—those hit by cars. But even if many of your students have never seen a real opossum, *Mother Opossum and Her Babies* is a delightful unit—its "pouch" is filled with creative and fun activities designed to develop science and math concepts appropriate for young students!

In addition to young students discovering many things about opossums, marsupials, and animal behavior in general, this unit interweaves a great deal of mathematics and language arts learning. Measurement is the key as children compare themselves to an adult opossum, and later to a newborn and young opossum. Nonstandard measurement is introduced when children measure mother opossum and later a young opossum they have made. An exploration of pockets in the classroom provides further opportunities to explore number and place value for older children. Overall, the unit gains added relevance by giving young children many opportunities to compare and contrast their own growth and development with that of the opossum.

The opossum plays dead,
whenever there's dread

Of an approaching stranger,
who may mean danger!

He lies very still,
and is quiet until

The danger's gone away,
so . . . off goes the opossum to play!

— *Linette Mace and her first grade class
Hubbard-Radcliffe Elementary,
Radcliffe, Iowa*

In **Activity 1: Getting to Know Opossums,** a drama introduces children to the nighttime behavior of an opossum searching for food. Children find out that an opossum depends in large part on its sense of smell. As they role-play opossums sniffing in the night, children become more aware of how much they also rely on their sense of smell as they sniff a variety of items with strong odors. Later the children, pretending to be opossums, are challenged to identify some mystery foods by smell. After making their guesses, the class joins together to enjoy a snack of "opossum foods." Comparison continues as the students measure themselves next to a life-size poster of a mother opossum, and compare their hands and feet to an opossum's paws. Mathematics is naturally integrated with life science as the children measure the mother opossum with nonstandard units of measurement (such as blocks, popsicle sticks, and other common objects). The students gain practice with estimating, counting, and, for older students, recording data.

In **Activity 2: Mother Opossum's Pouch,** one of the most intriguing characteristics of the opossum is investigated—her pouch. The development of the baby opossum presents excellent opportunities for more comparison, measurement, and counting activities in the context of dramatic play. Students guess what a mother opossum keeps in her fur-lined pouch. They are given tiny, pink, life-size, paper, baby opossums to hold and observe. Each child then puts on a paper-bag pouch that ties around the waist, in which to keep their "baby." As the children play, they pretend their babies are sleeping, nursing, and growing inside the pouch. This role playing stimulates your students' imaginations and allows them to learn interesting life science information in a playful way.

An unusual aspect of the opossum's pouch is that it contains 13 nipples. Pink beans, representing babies, are used in number activities to convey this interesting fact. Students count out 13 beans and glue them onto a drawing of the mother opossum's pouch. Older students play a game in which they estimate the number of beans in a partner's hand. Later, children count their own "pouches" (pockets) on their clothing and share with each other what they like to keep in their pockets. As a whole group they estimate and count the total number of pockets in the class using cubes as counters. The class creates a "train" of linking cubes and counts it by ones, tens, and fives, if appropriate.

In **Activity 3: Young Opossums,** students learn more about the growth and development of baby opossums. Each child makes a paper opossum with a pipe cleaner tail, giving it opened eyes and yarn fur to show that it is now ready to leave its mother's pouch. Children play with their young opossums outside, having them ride on their backs and climb through bushes. Through dramas, the children learn about the opossum's famous trick of "playing possum," then they act out this trick themselves. The children use their paper opossums for other active measuring activities, as they compare the length of objects found in the classroom with their opossums and sort the objects into three groups—those longer than, shorter than, and the same length as the opossum. A fun home activity invites families to find an object the same length as their young opossum and to bring it to school to share. Older children measure the opossum, as they did the mother opossum, but this time they use small items such as paper clips, pennies, or cubes. In the culminating session, the group creates a class book of drawings and writings about the opossum, and shares what they've learned with other classrooms, friends, and families.

The activities in this unit are designed for and field-tested by preschoolers, kindergartners and first graders nationwide. Much of the feedback from our field-test teachers indicates that the activities are appropriate for this wide range of abilities. We note certain activities as "Recommended for Preschool" or "Recommended for Kindergarten and First Grade" to give you an idea of this range. Yet, you know your children best, so be sure to read through the activities beforehand and make your selections accordingly.

Teachers have commented on the flexibility of the activities. Many teachers use the activities successfully in centers and learning stations, preferring to introduce the experience as a whole group, then sending students off for independent work. This format allows children to work at their own pace and ability, with concepts and skills reinforced over many days.

Teachers and children devised a number of modifications and suggestions for extending the learning in the classroom and at home. Look for these activities, storybooks, poems, and websites throughout. One great idea is to send a letter home at the beginning of the unit, apprising parents and families of the concepts and skills to be introduced. Families enjoy being a part of the classroom activities and, in the process, reinforce the concepts at home. We're sure you will come up with many adaptations of your own.

We hope you enjoy your study of this creature of the night. Perhaps it's the many engaging qualities of this unique animal that makes *Mother Opossum and Her Babies* appeal to children of all ages!

National Standards and

Not only does the development and parenting of the opossum's young provide children with an opportunity to explore this animal's intriguing life cycle, it makes an excellent venue to address key content science and mathematics standards.

The science standards are outlined in the *National Science Education Standards* from the National Research Council, while the mathematics standards are spelled out by the National Council of Teachers of Mathematics (NCTM) in its innovative 1989 document, *Curriculum and Evaluation Standards for School Mathematics*. That book set national standards for mathematics education to guide the curriculum and instruction of all children in grades K–12. These standards are further refined in the soon-to-be-published *Standards 2000.* An important change is the inclusion of standards for early childhood.

Mother Opossum and Her Babies addresses the following science standards (Content Standard C—Life Science for grades K–4):

The Characteristics of Organisms

- Each plant or animal has different structures that serve different functions in growth, survival, and reproduction.
- The behavior of individual organisms is influenced by internal cues (such as hunger) and by external cues (such as a change in the environment). Humans and other organisms have senses that help them detect internal and external cues.

Life Cycles of Organisms

- Plants and animals have life cycles that include being born, developing into adults, reproducing, and eventually dying. The details of this life cycle are different for different organisms.

Organisms and Their Environment

- An organism's patterns of behavior are related to the nature of that organism's environment, including the kinds and numbers of other organisms present, the availability of food and resources, and the physical characteristics of the environment.

Mother Opossum and Her Babies

Mother Opossum and Her Babies also includes activities that involve estimation, number, and measurement. In addition, as young students do the activities in the guide, they develop problem-solving skills, mathematical reasoning and communication skills, and see the connection between mathematics and their world. Estimation, number, and measurement are taught in connection with one another to promote the idea of "making sense" in mathematics—each topic is taught in connection to each other, and not in isolation.

The NCTM Standards related to mathematics content that this guide addresses include:

Estimation

- Exploring estimation strategies
- Recognizing when an estimate is appropriate
- Determining the reasonableness of results
- Applying estimation in working with quantities, measurement, computation, and problem solving

Number Sense and Numeration

- Constructing number meanings through real-world experiences and the use of physical materials
- Understanding our numeration system by relating, counting, grouping, and place value concepts
- Developing number sense
- Interpreting the multiple uses of numbers encountered in the real world

Measurement

- Understanding the attributes of length, capacity, weight, mass, area, volume, time, temperature, and angle
- Developing the process of measuring and concepts related to units of measurement
- Making and using estimates
- Making and using measurements in problem solving and everyday situations

Activity 1: Getting to Know Opossums

Overview

To begin this unit, the children watch a short drama about a hungry opossum that goes to houses on dark nights to look for food. In a partially darkened room, the youngsters investigate various items as they role-play opossums sniffing in the night. Later, they sniff and guess the identity of hidden foods and eat them as an opossum snack. Through these enactments of opossum behavior, children begin to see how an opossum's sense of smell and nocturnal activity help it to live successfully in its environment. The opportunity for the children to explore their own sense of smell encourages learning about themselves and how they compare to other living things.

Next, younger children discover they are not much longer than a mother opossum when they lie down on the floor next to a life-size drawing of a mother opossum to compare their length to hers. Children measure the mother opossum using common classroom items such as building blocks. They also compare their hands and feet to drawings of an opossum's paws.

The activities introduce the concepts of size, length, and measurement in a way that is relevant and engaging for youngsters.

These activities are easily adapted to a center or learning station format. Many teachers introduce the lessons as a whole group and then provide the materials for further exploration at a learning station or choice center.

Session 1: Introducing Opossums

What You Need

For the group

❏ 1 Mother Opossum poster
(The poster is at the back of the guide.)

❏ Gray and pink crayons, markers, or paints

❏ 1 piece of white paper 3" x 18" for making an opossum tail

❏ Scissors

❏ 1 piece of tape

❏ 1 small bowl of dry or canned cat food
    (Cat food with a strong odor that the children can smell works best.)

❏ 5–10 bowls or plates of strong smelling items such as soap, scented candles, toothpaste, fragrant flowers, half a lemon, mint leaves, parsley, garlic, cat food, coffee beans

Optional

❏ 1 toy opossum

❏ 5–10 lengths of yarn, each 6–10 feet long

❏ Journals for each student

Sample Letter to Families to Introduce Opossum Activities

Dear Families,

We are beginning a new science unit in our classroom about opossums. To help your child learn about the unique characteristics, behaviors, and the development of opossums, they will be actively involved in a series of projects and lessons. Dramas will be used to introduce the content information about opossums, and you may see your child perform some of the dramas at home.

Children will make a paper opossum at school, which they will use for a measurement activity at home with you. A letter with details of that project will be sent home at a later date.

As a part of this unit, there are many opportunities to make connections to mathematics and language arts. Your child will estimate, count, compare, and measure. All these activities will help develop their understanding of numbers and measurement. In addition, storybooks will be used to support each lesson. The children will develop their writing skills by recording their measurement results and creating pages for a class book.

For your information, opossums are part of the marsupial family of animals (pouched animals) that includes kangaroos and koala bears. As part of the unit, the children will have fun wearing a "pouch"—just like a mother opossum! Opossums are the only marsupials in North America and are widely found in the United States. They have the ability to survive in both city and country areas. Please share any information about opossums you have—from firsthand experiences to books with pictures.

The children are excited about opossums and we are looking forward to a unit filled with fun and learning.

Getting Ready

Anytime Before the Activity

1. Cut an opossum tail out of the white paper.

2. Send a letter home to parents describing the activities to come. A sample letter is on page 8.

3. If you decide to use journals to record information, make one for each child.

Immediately Before the Activity

1. Color or paint the Mother Opossum poster. The body should be gray and the nose pink.

2. Place the plates of "smell" items in different parts of the room, on the floor, low shelves, etc.

3. Place the cat food in the middle of the area where the children will watch The Opossum Drama.

4. Tape the paper tail onto yourself.

   Optional

   1. Hide the toy opossum nearby.

   2. If you like, have 6–10 feet of yarn stretched out from each plate along the floor to designate a "trail" for the children to follow.

The Opossum Drama

1. Gather the children in a half circle on the floor. Partially darken the room. You may want to use a lamp to represent moonlight. Pretend to be an opossum as you tell the following short story. If you have a toy opossum, use it to act out the drama.

   - On dark nights, a hungry animal comes out of the woods. It walks quietly around houses looking for cat food that is left outside. (Crawl around on the floor.)

   - This furry animal has gray fur, a long white tail, and a very pointed nose.

   - The animal sniffs and sniffs. (Have the children sniff with you.) Something smells good.

   - The animal finds a dish filled with cat food. The dish makes loud, clanking sounds as the animal pushes the dish around and eats the food. (Push the bowl with your mouth and pretend to eat the food.)

   - What do you think it is?

2. The children may mention a dog, cat, mouse, rat, or a variety of other animals. Acknowledge their answers as good guesses. Use their answers as a way to give more hints about this mysterious animal. For example:

   - If a child mentions a raccoon say, "That's a good guess. This animal comes out at night like a raccoon, but it doesn't have stripes on its tail and it doesn't have a mask."

- If a child mentions a mouse say, "That's a good guess. It looks like a mouse, but this animal is much bigger than a mouse. It's this big." (Use your hands to show the approximate size of an opossum.)
- If a child mentions a cat say, "That's a good guess. Its body is furry like a cat's, but its tail is different. Its tail doesn't have fur on it."

If possible, arrange to take a field trip to see a real opossum or to have one visit your classroom. Call your local nature center or wild animal rescue facility for information.

The Mother Opossum Poster

1. Show the Mother Opossum poster to the children and tell them it's the animal in the story. Ask, "What do you think it is?"

2. If the youngsters don't guess an opossum, tell them.

3. Ask questions to encourage the children to talk about opossums, such as:

   - Have you ever seen a real opossum? What was it doing?

   - What do you notice about this opossum?
     [It has fur, a long tail, and a pointed nose]

   - How many legs does it have?
     (Along with the children, count the opossum's legs.)

4. Have a child find the opossum's nose.
   Ask, "What do you use to smell?" *[My nose]*
   "What do you think an opossum uses to smell?" *[Its nose]*

Role Playing Opossums

1. Tell the children opossums have weak eyesight, but they have a very good sense of smell.

2. Tell the children they are going to pretend they are opossums sniffing around at night. Ask, "What do opossums use to help them find food when it is dark?" *[Their noses]*

3. Partially darken the room, and encourage the children to role-play opossums. They can crawl around on the floor and sniff the contents of the different bowls you placed earlier around the room. Remind them not to taste any of the things in the bowls.

   Optional
   Before they search for the food bowls, tell them to follow a "yarn trail" to find a bowl to sniff. They can go from one trail to another until they've found them all. (This may help with crowd control rather than 20 opossums randomly looking for food at the same time!)

4. Turn on the lights and gather the children back together. Encourage them to identify the items and describe the different smells.

   Which ones did they like?

   Which ones did they not like?

   Which one was their favorite?

Since young children love to dress up like animals, some teachers made paper tails for their students to wear during the role-play.

Some teachers had their children begin "Opossum Journals" by having them record the smells with drawings or words, and circle the ones they liked.

Session 2: Young Opossum Snacks

What You Need

For the whole group

❏ 1 sharp knife

For each group of six children

❏ 3 or more small, opaque, plastic containers, such as 8-ounce yogurt containers, with lids

❏ 3 or more types of food with strong, identifiable smells (such as orange, banana, peanut butter, pickles, tuna fish)

For each child and yourself

❏ A small serving of each of the foods you select to put in the scent containers

❏ 1 plate

Optional

❏ Several crackers (if you serve peanut butter as a snack)

Getting Ready

1. Divide the class into small groups of six or less. Make three scent containers for each group. All the groups can have the same foods in their containers, or you can make them different, and let the groups trade with each other. Alternatively, you can set up the containers at one place for the children to rotate through.

2. Use the containers with lids to make scent containers.

    a. Put a piece of food in each container.

    b. Use the knife to punch six or more small holes in each lid.

    c. Sniff the containers through the holes in the lids to make sure the scents are strong enough. You may need to shake the containers to release more of the scent.

3. Prepare a serving of each type of food used in the scent containers for the children to taste. Put the snacks in a place where the children can't see them.

To down play the similarity to Session 1, emphasize the "mystery" nature of this sniffing activity by telling the children that they are going to guess the foods hidden in the cups. Explain that these are foods that opossums (and people) like to eat!

Young Opossums Sniffing for Food

1. Gather the children in a circle on the floor. Ask them to recall the smelling activity from the previous session, "What do opossums use to help them find food?" *[Their noses]* Discuss how opossums have such a good sense of smell that they can tell what the food is without seeing it.

2. Tell them that they will try and guess what food is hidden in the containers just by its smell.

3. Partially darken the room, and pretend with the children that it is night and you are young opossums looking for food.

4. Give each small group of children three scent containers. Have the children take turns sniffing each scent container.

5. Encourage them to guess what each scent is. If they have trouble identifying the scents say, "Try to find one that smells like peanut butter (or orange, or banana)."

6. After all the children have a turn sniffing and guessing, let the students open the containers to see what's inside.

Eating Opossum Snacks

Give the children the snacks of foods that were used in the scent containers. Tell them these are foods that opossums and children might like to eat. Encourage the children to sniff their food like opossums do before eating it.

Going Further

1. Have children record their guesses with drawings or words, and identify their favorites.

2. Bring in other foods to smell in mystery containers and make a graph of their favorites.

3. Let children perform the opossum drama. Prepare a paper opossum tail for each child to wear.

Preparing the food for the scent containers and snack can be time consuming. If you make fewer containers for a learning center or use only one food for the snack, it will reduce your preparation time.

Session 3: Mother Opossum and Me

(For Preschoolers)

Kindergarten and First Grade teachers can conduct a shorter version of this session to introduce measuring mother opossum in Session 4. The children find it interesting that the drawing is full size and not that much shorter than themselves.

What You Need

For the group

❏ 1 Mother Opossum poster

Comparing Mother Opossum and Me

1. Show the Mother Opossum poster to the group.

2. Ask a child to point to the tip of the opossum's nose. Ask another child to point to the end of the opossum's tail.

3. Tell the children the poster is the same size as real mother opossums. Father opossums are a little bit bigger.

4. Ask, "Do you think you are longer or shorter than the mother opossum?"

5. Place the poster on the floor. Have a child lie next to the poster to make the comparison. (Be sure the child lines up the top of her head with the tip of the opossum's nose.)

6. Let all the children take turns comparing their lengths with the opossum drawing. Before each child takes a turn ask, "Do you think you are longer or shorter than the opossum?"

7. Leave the poster on the floor or at a central place so the children can continue to compare themselves with the mother opossum.

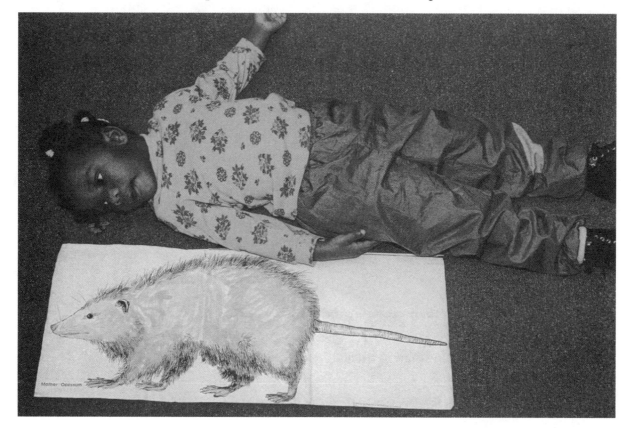

Session 4: Measuring Mother Opossum with Common Objects

What You Need

For yourself

- ❑ 1 Mother Opossum poster
- ❑ Enough blocks (such as building blocks) that are the same length to extend the entire length of the Mother Opossum poster
- ❑ Enough of another type of common classroom object of the same length (such as popsicle sticks or paper clips) to extend the entire length of the Mother Opossum poster
- ❑ 1 recording sheet

For the group

- ❑ Common classroom and real-world objects, such as popsicle sticks, paintbrushes, clothespins, toothpicks, plastic animals, or plastic spoons. You need enough of each item to extend the entire length of the Mother Opossum poster. Be sure all objects of the same type are the same length and do not roll easily.
- ❑ Baskets or containers to hold the objects

For each child or pair of children

- ❑ 1 copy of the Mother Opossum poster (these do not need to be colored)
- ❑ 1 marker and pencil for recording
- ❑ 1 recording sheet (or journals)
  (For Kindergarten and First Grade, use recording sheet on page 28)
  (For Preschool, use recording sheet on page 29)

Getting Ready

1. Duplicate the Mother Opossum poster as follows:
   - For preschoolers, make 1 copy for each child.
   - For kindergarten and first grade children, make one copy for each pair of children.
2. Prepare baskets or containers of materials for children to use for measuring. Have at least one basket/container of objects per child. Objects can be duplicated. For example, for a class of 20 children, materials could include:

   2 sets of **blocks** (all the same length)

   2 sets of **clothespins**

   2 sets of **paint brushes**

   2 sets of unsharpened **pencils**

   2 sets of **plastic animals** (same kind and size)

   2 sets of **large paper clips**

Consider the counting abilities of your students in selecting items for measuring. For preschoolers, use longer items that will measure the length of mother opossum with 12 items or less. For kindergartners and first graders, include some shorter items, such as paper clips or bottle caps, which will give them an opportunity to use larger numbers.

Also, select items that are easy to manipulate. Some younger children have trouble lining up objects that roll, such as straws, crayons, and paintbrushes.

When making copies of the Mother Opossum poster for your students, we found it easiest to fold the poster in half, duplicate the front half and the back half using tabloid paper (11" x 17"), and then glue or tape the two halves together. If your copier does not have tabloid paper, you can duplicate the poster using four sheets of letter paper (8 ½" x 11").

Some teachers said
band-aids and cotton
swabs (such as Q-tips)
are too tempting for
young children. They
want to play with the
items, opening and
using the band-aids
and putting the Q-tips
in their mouths and
ears.

2 sets of **popsicle sticks**

2 sets of **spoons** (same size)

2 sets of **index cards**

2 sets of **toothpicks**

3. To have the children record their measurements, select the method that matches the skills and abilities of your students.

   a. Have students use their journals or provide blank paper for students to trace the objects they use to measure. After they know how many of that item are needed to extend the entire length of the poster, they can record that number on or near the traced object.

   b. For preschoolers, use the recording sheet on page 28. Duplicate one per child. On this sheet, the measurement tool column is blank for children to draw the tool or have an adult write in the name of the tool they are using to measure mother opossum. It has a second column to record the actual measurements.

   c. For kindergarteners and first graders, use the recording sheet on page 29. Duplicate one per child. On this sheet, there are three columns: the measurement tool column, a column for their estimates, and a column for the actual number. Emphasize that estimates or guesses are not meant to be the actual number—often they are incorrect! However, as they practice making estimates, they will improve in their ability to come close to the actual number.

     Suggestion: Have children use crayons or pens to record so you can see their thinking—often children want to change their estimates to the actual number! Their estimates provide valuable assessment information.

     As you measure mother opossum with your class, demonstrate how to use the recording method you chose.

Accept guesses without
comment. Be aware that
some guesses may be high
or low. This estimation
provides an informal
assessment about develop-
ment of number sense.

Many times young children
enjoy giving really large
numbers, such as a hundred
or millions or even infinity.
Guide them to guesses that
are more realistic.

Recording Sheet

MEASURING TOOL	? PREDICTION ?	✓ ACTUAL ✓
BAND-AIDS		
BLOCKS		
CRAYONS		
PENCILS		

4. Decide how you want to set up the classroom for the "More Measuring" part of the activity. You many want to have the measuring tools in one location from which the children take one type of tool at a time and work at their tables, or you may want to have a learning center with the various tools, that the children rotate through.

Measuring Mother Opossum Together with Blocks

1. Put the poster of the mother opossum on the floor and gather the children around it.

2. Tell the children the class is going to measure mother opossum together. Have the children find the tip of the opossum's nose and the end of her tail.

3. Hold up a block and tell the children you'll use blocks to measure mother opossum. Line one block up with the tip of the opossum's nose.

4. Ask the children to guess how many blocks it will take to go across the mother opossum from the tip of her nose all the way to the end of her tail. Have them tell their guesses to a person sitting next to them, then ask for a few guesses.

5. As the children count with you, continue to put the blocks down, one at a time, in a straight line, with each block touching the next, until you reach halfway across the length of the opossum.

6. Ask the children if they have new guesses about how many blocks it will take to go across mother opossum. Have them tell their guesses to a person sitting near them. Next, listen to their revised guesses to see if and how their guesses change with more information.

7. Continue placing blocks across mother opossum until the length of her body is covered. Ask, "How many blocks did it take to go across the length of the mother opossum?" Have the children count the blocks as you point to each one.

Measuring Mother Opossum Together with Other Objects

1. Choose another object, besides blocks, that the children are familiar with to measure the mother opossum, such as popsicle sticks or spoons. Hold up the item and have the children identify it.

2. Clear all the blocks and keep one available. Place one spoon on mother opossum and line it up with the tip of her nose. Ask the children if they think, using spoons, it will take more spoons than blocks, fewer spoons than blocks, or the same number of spoons as blocks. Depending on their responses, you may want to place a block next to the spoon for a concrete comparison.

3. Ask children to estimate how many spoons it will take to go across the entire length of mother opossum. Again, have them share their initial estimates with a person sitting near them. Listen to a few estimates.

4. As the children count with you, place the spoons down, one at a time, in a straight line with each end touching the next until half the length of the mother opossum is covered.

5. Ask if anyone wants to revise their original estimate. Listen to their new predictions.

6. Finish measuring the length of mother opossum and count the spoons again with the children. How close did their estimates come to the actual number? How did the number of spoons compare to the blocks?

If you are going to have your children record when they do the "More Measuring" part of the activity, be sure you model the recording method they will use as you measure the second time with the spoons or other objects you use.

Some items will not cover the length of mother opossum exactly. Let your children decide what to do in such cases. Some may use words like "almost" and "a little bit more" to describe the measurement. This is fine. Others may use "half" or a "quarter" to describe the measurement. You do not need to teach fractions, but you can explain the terms as appropriate.

More Measuring

1. Tell the children they now get to measure mother opossum for themselves.
   Note: With preschoolers, give each child her own Mother Opossum poster to measure. Kindergartners and first graders can work in pairs. Model how to work with a partner if necessary.

2. Show them the objects they can choose to measure mother opossum, including the ones you modeled. Tell them to be sure to predict how many of each object it will take to measure the length, **before** they measure and count.

3. Remind them to select only one container of objects at a time to measure with. When they have completed measuring with that object, they should return it and select another.

4. Distribute Mother Opossum posters to the children and send them to the work areas to begin. Circulate as they work and assist children as necessary. Remind them to make predictions before they begin measuring.

5. As you observe the children at work, listen to their predictions and informally assess their measurement skills. For those children who are recording, check to see that they are doing it accurately and assist them as necessary.

6. At the end of the work period, have the children meet again in the circle area to discuss what they discovered. If they recorded, have them bring their recording sheets or journals.

7. Compare the results of their measurements. Discuss why some children may have gotten different numbers than others using the same object. You may want to measure mother opossum as a group if there are large discrepancies in the measurements.

8. Make the poster and measurement materials available at a learning center for students to revisit this activity. You may want to add a new measuring tool.

Alternatively, set up learning centers with a Mother Opossum poster and one type of measuring material so children can circulate to each center with their recording sheet to take measurements.

In these early years, it is common for children to reverse the direction of numerals as well as the digits in the numbers as they record. It is still important that they record their findings! Provide models of numerals for them as reference. Ask questions if their recorded numbers seem way off.

In one classroom where parents participated, the children chose to measure a dad!

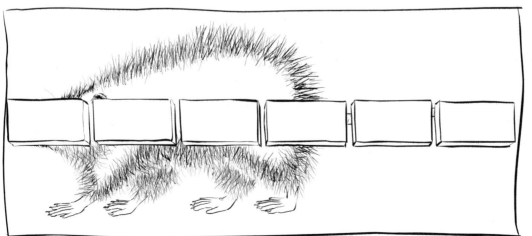

Going Further

1. Measuring Each Other

Have a child lie down. Have the other children predict how many objects long the child is. (Be sure to use a long object for this measurement with preschoolers.) Use the objects to measure him halfway, placing them alongside the length of the boy. Have children revise their original estimates. Finish measuring him and count the total number of objects he is long. Ask if anyone thinks they are about the same length. Have them lie down next to the line of objects to compare themselves.

> How many children are the same size?
>
> How many shorter?
>
> How many longer?

Kindergarten and first grade children can work with a partner to measure each other with another object. Have them predict first and then measure. Compare class results.

2. Measuring Classroom Objects

Encourage children to explore measurement on their own. Have them choose something in the classroom environment to measure. Decide which nonstandard unit of measurement they will use to measure. Preferably choose from the objects that were used to measure the Mother Opossum poster. Assist the children as necessary as they place the objects next to each other in a line and count the total number. Record the results on paper or in journals. Have the children report on their findings.

> Which item was the longest?
>
> Which item was the shortest?
>
> How many items are about the same length?

3. Measuring Things From Home

Decide on a category of items for children to bring from home, such as toy vehicles, toy animals, or something made of fabric. Have each child bring in one item. (Be sure to have a few extra items on hand for children who forget to bring one.)

Choose a unit of measurement to determine the length of each item. Set up a center with the measurement tool. Have children predict, then measure, to determine the length of the item. When everyone has had a turn, meet as a group to share results.

> Which item was shortest?
>
> How many were the same size?
>
> Which was the longest?
>
> Finally, line up the items from shortest to longest.

Session 5: An Opossum's Paws

What You Need

For the group

- ❏ 1 Mother Opossum poster
- ❏ 1 container of finger-paint, or poster paint (any color)
- ❏ 1 tray
- ❏ Enough newspaper to cover the work table

If the children prefer, they can trace their hands and feet next to the paw prints. They may like to add features, such as fingernails and toenails.

For each child and yourself

- ❏ 1 Opossum Front Paw student data sheet (see page 26)
- ❏ 1 Opossum Back Paw student data sheet (see page 27)

Getting Ready

Anytime Before the Activity

1. Copy one Opossum Front Paw student data sheet and one Opossum Back Paw student data sheet for each child and yourself.

Immediately Before the Activity

1. Spread newspaper on the table.
2. Spread the finger-paint on the tray, and put the tray on the table.

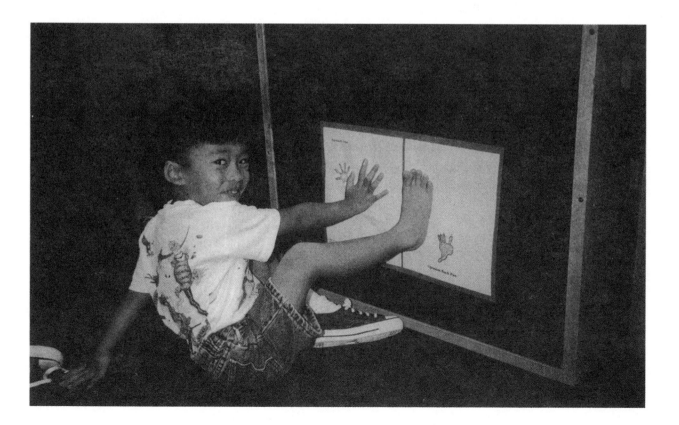

Comparing Front Paws and Hands

1. Gather the children on the floor in a circle. Show them the Mother Opossum poster.

2. Say, "Guess what an opossum's hands and feet are called." If the children don't know, tell them they are called *paws*.

3. Have a child point to a paw on the Mother Opossum poster. Along with the children, count the paws.

4. Give each child a copy of the Opossum Front Paw drawing.

5. Say, "Guess what this picture shows." If the children don't know, tell them the picture shows an opossum's paw.

6. Have each child put his or her right hand on the paper next to the drawing of the opossum paw. Ask questions such as these to encourage the children to look at the drawing and compare it with their own hands.

   - "Is your hand bigger or smaller than the opossum paw?"
   - "How many fingers do you have?"
     Along with the children, count the fingers on one hand.
   - "How many toes does the opossum have?"
     Have the youngsters count the toes on the opossum paw drawing.
   - "How is the opossum paw like your hand?"
   - "How is the opossum paw different from your hand?"
   - "What are some things you do with your hands?"
   - "What are some things you do with your feet?"
   - "How do you think an opossum uses its paws?"
     [*To walk, climb trees, and to hold onto things*]

Paw Prints Meet Hand Prints

1. Have each child, one at a time, place a hand into the paint on the tray.

2. Have the child press her hand with paint next to the opossum front paw drawing on the sheet of paper.

3. Use the questions above to encourage the children to compare their hand prints with the opossum's paw.

Comparing Back Paws and Feet

1. Give each child a copy of the Opossum Back Paw drawing.

2. Say, "Guess what this picture shows." If the children don't know, tell them the picture shows an opossum's back paw.

3. Have the children take off their shoes and socks and compare their feet to the opossum's back paw.

4. Ask similar questions to the ones you asked about the opossum's front paw—of course these questions are about feet—including,

> "Look at the opossum's back paw and your hand. How are they the same? How are they different?"

> "How do you think the opossum uses its "thumb" to help it climb?"

Optional

Have each child trace their foot or make a footprint with paint next to the Opossum Back Paw drawing. Making foot prints is a messy activity and requires extra adult assistance. Be sure to have a tub of soapy water ready for the children to immediately wash their feet.

Going Further

1. Compare the Mother Opossum poster and the paw print posters with a live adult cat, rat, or mouse. Compare their body and paws to those of an opossum.

2. Feed the visiting animals, and encourage the children to notice if the animals sniff the food and if they hold it in their paws.

3. Read *Animal Tracks* by Arthur Dorros (see the Literature Connections section on page 74) or other track books (see the Resources section on page 71), and discuss how tracks reveal information about an animal's behavior.

   Can they guess what animals left the tracks?

   What were they doing?

   Where were they going?

4. Go outside and see if you can find any animal prints.

5. With masking tape, tape each child's thumb to the side of their hand. Let them try to pick up things, read a book, or play with toys. What is it like to not have the use of your thumb? Relate their experience back to the help-fulness of the thumb for people and opossums.

6. Place a 2 x 4 inch board (6 feet long or more) on the floor and let the children "climb" along the board using their hands and feet. Also, have them hang on "monkey" bars using their hands. Discuss how our fingers, toes, and especially our thumbs help us to climb, grip, and keep our balance.

7. Continue to add to each child's "Opossum Journal" with drawings, recording sheets, stories, and paw prints.

OPOSSUMS

In the Fall, opossums get ready for Winter. The opossum is storing berries. After gathering the berries he went back into his house.
Jackson

Hillary Lee

First, the opossum found his home in a tree. Then he got all the berries and ate them before Winter came.
The End   Kolton

Opossum Front Paw

©1999 by The Regents of the University of California
LHS GEMS—*Mother Opossum and Her Babies*

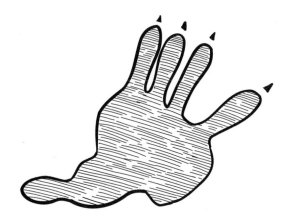

Opossum
Back
Paw

©1999 by The Regents of the University of California
LHS GEMS—*Mother Opossum and Her Babies*

RECORDING SHEET

MEASURING TOOL	✓ ACTUAL ✓

RECORDING SHEET

MEASURING TOOL	? PREDICTION ?	✓ ACTUAL ✓

Activity 2: Mother Opossum's Pouch

Overview

The unusual birth, growth, and parenting of young opossums offers a great opportunity for children to compare an opossum's early beginnings to their own. The children hold life-size drawings of newborn opossums and use these tiny babies for observation activities. The youngsters learn that a mother opossum has a pouch, and her babies stay in the pouch where they are warm and safe. The children watch with delight as the tiny paper opossums "climb" into a paper-bag pouch. The boys and girls pretend the babies drink milk from nipples in the pouch and grow bigger.

Later, the children use beans to represent the tiny baby opossums to play guessing and counting games. This is an opportunity to review the actual number of baby opossums, which can range from 1 to 13, that a mother opossum could have in her pouch.

Learning about the opossum's pouch naturally leads to making observations about pockets. Children are asked what they know about pockets, and about other animals that have pouches. Kindergartners and first graders estimate how many pockets the group has all together. They use unifix cubes as a tool to help them find out the answer, practicing one-to-one correspondence, counting by ones, and then counting by tens and fives.

Again, many of the activities lend themselves to a learning station presentation or as follow-up experiences at home.

Session 1: What's Inside the Pouch?

What You Need

For the group

- ❑ 1 Newborn Opossums sheet (see page 45)
- ❑ 1 pink crayon, or 9" x 11" pink paper
- ❑ 1 pair of scissors
- ❑ 1 roll of masking tape
- ❑ 1 hole punch

For each child and yourself

- ❑ 1 paper lunch bag
- ❑ 1 piece of string, yarn, or elastic (from a fabric store) long enough to tie around child's waist (make a longer piece for yourself)
- ❑ 1 paper newborn opossum

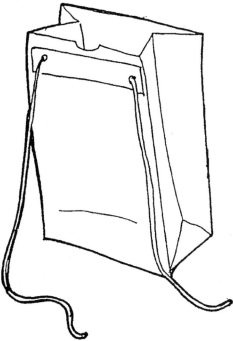

Getting Ready

Anytime Before the Activity

1. Copy the Newborn Opossums sheet onto white or pink paper. If you use white paper, color the newborns pink. Cut out one newborn opossum baby for each child in your group.

2. Follow the steps below to make a paper-bag pouch for each child and yourself.

   a. Cut a 5" piece of masking tape and tape it across one side of each bag near the top for reinforcement.

   b. Cut or punch two holes about 4" apart on the tape.

   c. Thread the piece of string through the holes.

Pockets

1. Ask questions to encourage the children to talk about pockets, such as:

   - Who has a pocket?
   - Find all the pockets on your clothes. How many do you have?
   - What do you keep in your pockets?
   - Why do people's clothes have pockets?

2. Give the children plenty of time to talk about their pockets, the usefulness of pockets, and to find the pockets on their clothes.

3. Ask if they know of any animals that have a "pocket" or pouch. Many may mention kangaroos or koalas. Ask, what do they use their pouch for?

Teeny, Tiny, Newborn Opossums

1. Gather the children in a circle on the floor. Tell them all mother opossums have a pouch. The pouch is like a big pocket.

2. Tie the paper-bag pouch around your waist. Pretend you are a mother opossum with a pouch.

3. Ask, "What do you think mother opossums keep in their pouches?" *[Babies]*

4. Ask, "How big do you think baby opossums are when they are first born?" Have the children use their hands to show you.

5. Give each child a paper newborn opossum to hold. Tell the group the tiny paper opossums are the same size as real newborn opossums.

6. Ask questions to encourage the youngsters to observe the newborns, such as:

   - What do you notice about the newborn opossums?
   - Does the tiny baby fit in your hand?
   - Does the tiny baby fit on your thumb nail?
   - What color is the tiny baby? *[Pink]*

- Can you find its little legs and tiny tail?
- Do you see its eyes?
- Are the eyes open or closed? *[Closed]*

7. Tell the group that when real opossums are born their eyes are closed and they have no fur, only pink skin.

8. Tell the youngsters the tiny opossums crawl into the mother's pouch right after they are born. Have the children crawl their paper babies one at a time into your pouch, and count them as they go.

9. Tell the group a real mother opossum keeps many babies in her pouch. Ask, "What do you think the babies do in the pouch?" Accept the children's answers. As needed, explain that the tiny opossums stay in the pouch where they are warm and safe. They find a nipple, drink milk from it, and grow bigger.

Wearing Opossum Pouches

1. Tie a pouch onto each child who wants to wear one.

2. Allow time for the boys and girls to play freely with their newborn opossums and their pouches. Encourage the students to nurture the tiny babies by keeping them in their pouches where they are warm and safe.

3. Let the children wear their paper bag pouches during play time to emphasize that the tiny opossums stay in their mother's pouch everywhere she goes. They can climb on the climbing structure or in trees, pretending to be opossums. They also can wear their pouches during snack time, lunch or story time.

One teacher told us some of the boys in her class did not want to pretend to be mother opossums, so they pretended to be father opossums with pouches instead.

Of course, male opossums don't have pouches. But you could mention that male sea horses have pouches where they keep the baby sea horses safe.

Invite upper grade "buddies" to come in and help tie on pouches and participate in activities using beans as babies in future sessions.

4. Ask questions to review what the children learned about newborn opossums.

- Why do you think the babies need to stay in the pouch when they are so little? *[To be safe, to have a warm place to sleep and eat]*
- Why do you think the babies would get cold if they were out of the pouch? *[They don't have fur to keep them warm]*
- What do the tiny newborns drink when they are in their mother's pouch? *[Milk]*
- If the newborn's eyes are closed, how do you think they find the pouch? *[They use their sense of smell]*

5. Save the children's pouches to use in the following session.

Going Further

1. Study other animals that have pouches such as kangaroos, koalas, sea horses, and pill bugs. Use books, pictures, drawings, and toy animals to compare their "babies in a pouch" behavior.

2. Read the books *Katy No Pocket* by Emmy Payne and *A Pocket For Corduroy* by Don Freeman. (See the Literature Connections section on pages 74–75.)

3. Introduce marsupials and study other marsupials from Australia.

4. Have children design and make their own pouches out of a folded paper plate, yarn/string, furry material, and pieces of straws for nipples.

5. Have children role-play being inside of a warm pouch by snuggling with others under a large blanket.

Session 2: How Many Babies in Mother Opossum's Pouch?

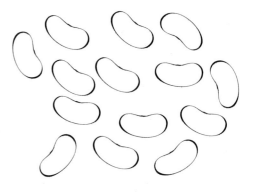

What You Need

For each child and yourself

- ❏ 13 pink beans
- ❏ 1 Mother Opossum's Pouch sheet (see page 42)
- ❏ 1 paper-bag pouch made in Session 1
- ❏ 1 pencil or marker
- ❏ White glue
- ❏ 1 cup

Getting Ready

Anytime Before the Activity

Make a copy of the Mother Opossum's Pouch sheet for each child

Immediately Before the Activity

1. Fill each cup with 13 beans.
2. Tie the pouch around your waist and put 13 beans inside.

How Many Babies Are in the Pouch?

1. Show the children a pink bean. Tell them the bean is the same size as a real newborn baby opossum. Ask the children where the baby opossums crawl after they are born. *[Into the mother opossum's pouch]*

2. Tell the children a mother opossum has a lot of nipples to nurse her babies. Show the Mother Opossum's Pouch sheet to the children. Count the nipples with the children. Tell them that since she has 13 nipples, 13 is the most babies a mother can nurse in her pouch.

3. Write your name on your Mother Opossum's Pouch sheet. Ask the children to count with you as you glue one bean on to each nipple on the sheet. Ask the children to tell you how to write the number 13 on the sheet, and then write it.

4. Send the children to the tables to glue 13 beans onto their Mother Opossum's Pouch sheets. Remind them to write their names on their sheets. Have the number 13 written where children can see it and copy it if necessary.

Counting Bean Babies

1. Tie a pouch around your waist and put 13 beans inside.

2. Tell the children they are going to play a game with beans. In the game, they will pretend the beans are baby opossums.

3. Ask the children to pretend you are a mother opossum. Tell them you have some "babies" in your pouch. Ask, "What is the largest number of babies a mother opossum can have in her pouch?" [13]

4. Put your hand into the pouch and take out a handful of "babies." Have the children look at your handful of "babies."

5. Line up the "babies" and count them with the children.

6. Return the beans to your pouch. Again, put your hand into the pouch and take out some "babies." Have the children look at the handful of beans. Line up the bean "babies" and count them with the children.

7. As a whole group or at a central place, give the children their pouches to wear and give each child 13 beans to put inside. Let the boys and girls wear their pouches and play the game with a partner. They can take turns taking a handful of "babies" out of their pouch. Then, together, the partners can line the beans up and count them.

Estimating Bean Babies

Recommended for Kindergarten and First Grade

With older children, repeat the game, but this time have children estimate the number in a handful before counting. For a more challenging game, place more beans in the pouch.

Remind students that estimating the number of beans is not a contest! Though there may be some "closer" estimates, none are "wrong." With practice everyone can become a skilled estimator.

Going Further

1. Send another Mother Opossum's Pouch sheet home with beans as a family activity.

2. Prepare enough paper newborns so children can glue them on the Mother Opossum's Pouch sheet instead of beans.

Session 3: Counting Our Pockets

Recommended for Kindergarten and First Grade

What You Need

For the group

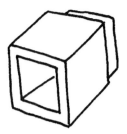

❏ 10 unifix cubes per child, any color

❏ 5 baskets or more (depending on the number of children in class)

Getting Ready

1. For a class of 20 children, you will need 200 unifix cubes. Fill each of five baskets with 40 unifix cubes. Place the baskets in easily accessible locations so that groups of four children can take cubes and make "trains" of cubes.

2. Be sure to wear clothing with at least one pocket!

Estimating Pockets

1. Pose the following questions to your students to launch this activity.

   • Why do opossums have pouches? *[To hold their babies]*

   • Do we have pouches like opossums?

   • What do we have instead of pouches? *[Pockets]*

   • Who has a pocket on their clothes today?

2. Have the children find all the pockets they have on their clothing with a partner, so they are sure to locate all their pockets. Have each child find out how many pockets they have. Then, have the youngsters estimate the total number of pockets in the classroom by asking:

- How many pockets do you think everyone is wearing?

Give the children a chance to discuss this question with a partner. As they share their estimates with the group, record the numbers without recording names.

Counting Pockets

1. Ask for a method to find out how many pockets everyone has on their clothes and listen to the children's ideas. Tell your students they will use unifix cubes as a tool to help count the total number of pockets.

2. Model how to use the cubes. Put one cube in each of your pockets. Let them know they will do the same.

3. Have the baskets of unifix cubes where groups of four children have easy access to them. After everyone has put one cube in each of their pockets, collect the extra baskets of cubes.

4. Remove the unifix cubes from your pockets. Model how to make a "train" with them by linking them together.

5. Have students make trains with their cubes. Have the children count the number of cubes in their trains, and compare their trains with the trains of children near them.

>   Are some trains the same length?
>
>   Which train is the shortest?
>
>   Which train is the longest?

6. Make statements about the trains so the class can continue to compare their lengths. Have the youngsters hold up their trains if they agree with a statement you make. Children who do not have a train (those that had no pockets) can raise their hands for all the statements that apply to them. Use the following statements or others appropriate for your children:

   • If you have zero cubes in your train, hold up your hand.

   • If you have fewer than three pockets, hold up your train or hand.

   • If you have two cubes in your train, hold it up.

   • Hold up your train if you have more than four cubes in it.

   • Hold up your train if it has an even number of cubes.

   • Hold up your train if it has an odd number of cubes.

   • Hold up your train if you think it has the more cubes than anyone else's train. Compare those trains.

   • How many cubes are in the longest train(s)? Count the cubes.

7. Remind the children that the cubes represent the number of pockets they have. Tell them, in order to find the total number of pockets everyone has, they are going to link all the trains of cubes together.

8. Take your train and have a child add his train to it. Continue, child by child, until all the trains are linked together. Encourage the children to make new estimates about the total number of pockets as the large class train is created. Stop, at least once, and have them revise their estimates based on more information (the added cubes). At this point allow the children to whisper their new estimate to a friend.

9. Once the long train is finished with the class, count by ones up to the total number of cubes in the train. Ask your children how to record the number and write it near their estimates. Pose questions such as:

>   Were any of the original estimates close to the actual number?
>
>   Were their revised estimations close to the actual number?
>
>   Did it help to revise their estimates?
>
>   Why?
>
>   What do the cubes represent? *[Pockets]*

Counting Pockets Other Than by Ones

Before you count using an alternate method, ask the children how many cubes there will be. Their responses will provide an insight into their concept of number.

1. Ask if there are any other ways to count the total number other than counting by 1's. Depending on their response, use any appropriate strategy to count the cubes again, such as counting by 2's, 5's, and 10's.

2. Use the strategy of counting by 10's, to model a technique for counting.

   - Start at one end of the train, count 10 cubes. Break off that "10-train." Hold up the train and be sure everyone knows it is 10 cubes long.

   - Continue from the same end of the long train. Line up the "10-train" next to the long train and break off another "10-train" from the long train.

   - Continue to use the "10-train" to break the large train into additional "10-trains." When there are no more trains of 10, you will either have no cubes left or a train with less than 10 cubes. If the train has less than 10 cubes, break it apart into "ones."

   - Count the total number of cubes by tens and ones.

Children really enjoy this activity! Many get into finding and wearing clothes that have lots of pockets to increase the class total.

3. If appropriate, count the train again by using a "5-train" to measure trains of five.

4. Tell the children they will do the activity again the next day to see if the number of pockets changes. Have them guess: Will there be the same number, more, or fewer pockets?

Going Further

1. Continue to estimate, count, and record the number of pockets in the class over several days. Add, or use a calculator, to get the total after a week.

2. Have each child create a page for a "What Is In My Pocket?" class book. Make a double-sided sheet using pages 43–44 or the perforated pages in the back of the guide. Duplicate enough copies so each child has one double-sided sheet. Fold each sheet over to the thick line before distributing the sheets to the children.

   Have each child draw a picture of what is in their pocket on the inside flap of their sheet. Depending on your children's skills, have them also write the word(s) for what is in their pocket and their names.

   Using construction paper, make a front and back cover for the class book. Cut the construction paper so it covers the folded sheets—approximately 8½ x 6 inches. Stack the completed student sheets (still folded). Staple a cover to the front and back of the stack above the folded-over area.

   Read the book to the class. Give children an opportunity to take the book home to share with their families.

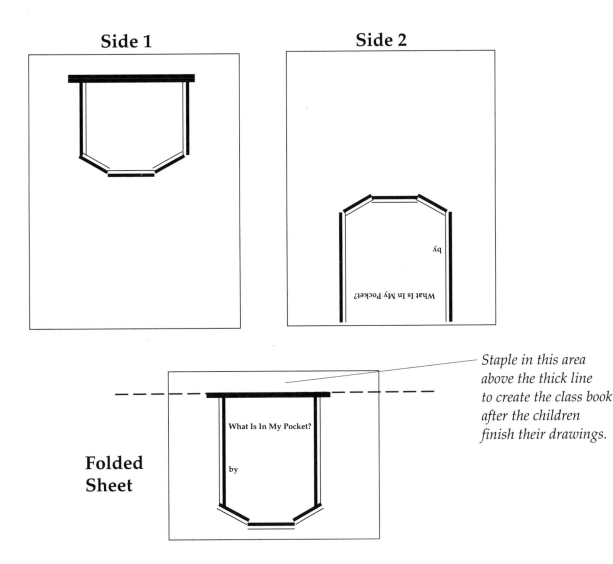

Side 1   Side 2

What Is In My Pocket?   by

Folded Sheet

What Is In My Pocket?   by

Staple in this area above the thick line to create the class book after the children finish their drawings.

Mother Opossum's Pouch

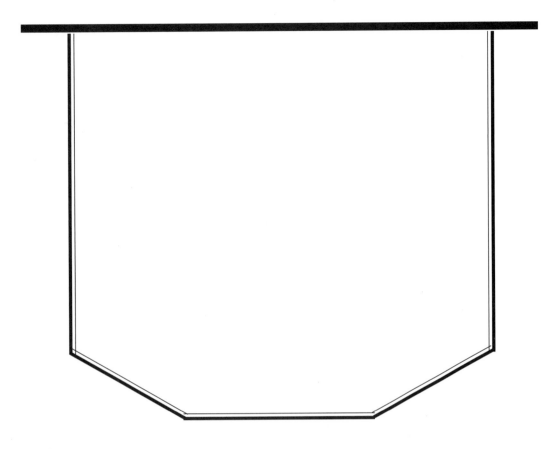

©1999 by The Regents of the University of California
LHS GEMS—*Mother Opossum and Her Babies*

What Is In My Pocket?

by _____

©1999 by The Regents of the University of California
LHS GEMS—*Mother Opossum and Her Babies*

Activity 3: Young Opossums

Overview

Exploring the growth and development of opossums continues in this activity. As baby opossums become older, they grow fur, and one day, crawl out of their mother's pouch and ride around on her back. In Session 1, the children make paper opossums. As the youngsters draw eyes and glue yarn onto their opossums, they pretend the babies are opening their eyes for the first time and are growing fur. The children use their paper opossums to act out opossum development from helpless babies to very active young opossums.

Outdoors, the youngsters pretend they are mother opossums. They carry the young opossums around in their paper-bag pouches and on their backs. The children make their opossums climb in bushes and roll over on their backs when a toy dog appears. The dramatic play and role-play reinforces the concept that opossums have a variety of behaviors and characteristics to help them survive in the wild.

Math investigations also continue as preschool children compare the length of their young opossums to the lengths of familiar objects in their classroom. Later, they go on a measurement hunt at home and bring back an item that is the same length as their young opossum. Kindergartners and first graders revisit measurement by measuring their young opossum with small classroom items.

In the culminating session, the children create a class book of drawings and stories that describe their in-depth learning about mother opossum and her babies.

A Perfect Pocket for an Opossum!

written and illustrated by:

KKELCEY ALEX KoITON BRIAN wiley Mi leon AL SARAH

Session 1: Little Opossums Grow Up

What You Need

For the whole group

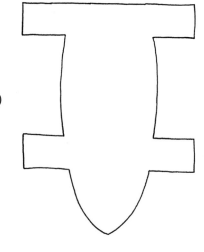

- ❏ 1 Mother Opossum poster
- ❏ 1 paper newborn opossum made in Activity 2: Mother Opossum's Pouch
- ❏ 1 Young Opossum Pattern (see page 62)
- ❏ 1 Young Opossum poster (see page 63)
- ❏ 1 pair of scissors
- ❏ 1 tray
- ❏ 1 container of white glue
- ❏ Enough newspaper to cover the tray and the work table

Although a bit more preparation time is needed to secure the pipe cleaner tails, it is definitely worth it. The flexible pipe cleaner simulates the prehensile tail of the opossum. This specialized tail allows the opossum to grasp and hold onto things much like another hand. Monkeys, too, have a prehensile tail.

For each child and yourself

- ❏ 1 4½" x 6" sheet of white construction paper, or a manila folder
- ❏ 1 9" white or gray pipe cleaner (for an opossum tail)
- ❏ Small pieces of white and/or gray yarn
- ❏ 1 paper-bag pouch made in Activity 2: Mother Opossum's Pouch
- ❏ 1 black marker
- ❏ 1 pink crayon or marker
- ❏ 1 paper clip

Getting Ready

Anytime Before the Activity

1. Use the Young Opossum Pattern to cut out one opossum from the white construction paper or manila folder for each child and yourself.

2. Cut the yarn into 1"– 2" pieces to represent fur.

3. Glue one pipe cleaner "tail" to each paper opossum. Be sure to use plenty of glue and gently clip the paper clip over the glued down end of the pipe cleaner to hold it securely. Place a layer of newspaper and books on top of the "opossums" for added pressure. It's best to make small stacks. Allow several hours for the glue to dry before you move the opossums. They may stick together slightly but should separate.

4. If the children are unable to write their own names, write each child's name on an opossum.

Optional

> If your children will be designing and making their own opossums, gather the materials they will need at a work area.

Immediately Before the Activity

1. Spread newspaper on the tray and the work table. Put a handful of yarn (to be used as "fur"), a black marker, and a container of glue on a tray in the discussion area and at each child's work area.

2. Put all the young opossum cutouts in one paper-bag pouch, and place the pouch near the discussion area.

Making Young Opossums

1. Show the Mother Opossum poster to the children. Encourage the youngsters to talk about what they like about opossums and some of the things they have learned about opossums.

2. Tie the paper-bag pouch around your waist and pretend you are a mother opossum with a pouch. Say,

   "The mother opossum has something in her pouch. Guess what it is?"
   [A baby opossum]

3. Ask, "Do you remember what the newborn opossum looks like?" "How big is it?" Show the paper newborn opossum to the children.

4. Say, "Let's pretend the tiny, newborn opossum drank a lot of milk and grew. Let's see what it looks like now."

5. "Climb" a young opossum out of your pouch and ask, "How is this opossum different from the newborn?"
   [It's much bigger and its tail is longer]

6. Say, "Let's pretend this baby's eyes are just beginning to open for the first time." Draw two eyes on the opossum.

7. Ask, "What else does the opossum need?" Draw a pink nose, whiskers, two ears on its head, and five claws on each paw as the children suggest them.

8. Turn the opossum over, and write your name.

9. Turn the opossum over again and spread glue over its back, legs, and head. While you place the yarn on top of the glue say, "Let's pretend the baby is growing fur."

Even if your children are creating their own opossums, it is still a valuable experience to conduct a modified whole-group introduction to steps 1–11. Discussing the changes taking place in the growing opossum as well as reviewing the body structures will help your children design a meaningful "opossum" to be used in future dramas and stories.

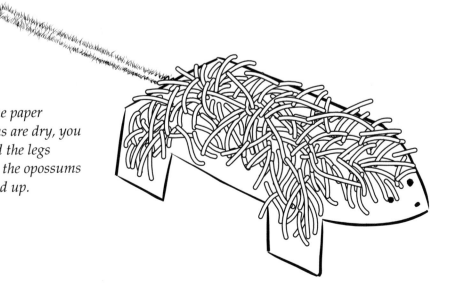

When the paper opossums are dry, you can bend the legs down so the opossums can stand up.

10. Say, "I feel more young opossums wiggling around in the pouch." "Climb" one young opossum at a time out of the pouch and give one to each child.

11. Tell the children to pretend their young opossums are opening their eyes for the first time and are growing fur. Have the children go to the tables to make their opossums. If the youngsters can write their own names, make sure they do it before they glue on the fur.

12. Show the Young Opossum poster to the students. Have them compare it to their paper opossums and to the Mother Opossum poster.

Session 2: A Young Opossum's Tricks

What You Need

For the group

❑ 1 toy dog or other opossum predator

❑ 1 toy frog or cricket

❑ 1 toy worm, or a 2" piece of brown yarn

❑ 1 nut

❑ Extra opossum food (toy frogs, crickets, worms, nuts)

❑ 1 tray

❑ Enough newspaper to cover the tray and the work table

Optional

❑ 1 cardboard tree described on pages 18–22 in the *Tree Homes* GEMS Teacher's Guide. The cardboard tree is made out of three cardboard boxes (as the trunk) and three or more cardboard tubes (as the branches). You can cut out leaf shapes from green paper to glue onto the branches as leaves. The cardboard is then painted with brown poster paint.

For each child and yourself

❑ 1 paper-bag pouch made in Activity 2: Mother Opossum's Pouch

❑ 1 young opossum made in Session 1

Optional

❑ 1 pair of scissors

Getting Ready

Immediately Before the Activity

Find an outdoor area with bushes and a place for the children to sit so they can watch and participate in the opossum dramas.

Riding on Mother Opossum's Back

1. Put on your pouch with your young opossum inside and pretend you are a mother opossum.

2. Encourage the children to pretend they are mother opossums with a pouch. Tie a paper-bag pouch around each child who wants to wear one.

3. Give the children their paper opossums to put in their pouches.

4. Tell the children that when opossums get bigger and no longer fit in the pouch, the mother has another way to carry her young. Let the children guess how she carries them.

5. If the children don't know, say "Let's find out where the young opossums ride when they get too big to ride in their mother's pouch."

6. "Crawl" the young opossum out of your pouch and onto your back. Walk or crawl with the opossum on your back. Ask the children how they think the young opossum holds onto its mother. *[With its claws]*

7. Take the group outside. Let the boys and girls play with their opossums—putting them in, and taking them out of, their pouches. Have the children ride the opossums around on their backs.

Although opossums are often described as hanging by their tails, they do not generally do so. They use their flexible tail for gripping, holding, and for balance. Some young opossums may grip a branch with their tail to lower themselves to another branch, but they do not hang suspended for any length of time.

Young Opossums in Trees

1. Take the group to the area with bushes. Let the youngsters carry the young opossums in their pouches as they walk to the bushes.

2. Gather the children on the ground near a bush. Tell the group the young opossums are getting bigger and are now old enough to climb trees.

3. Make your young opossum climb out of your pouch and climb on to a branch of the bush. Show how they can use their tail to grip onto a branch. Can they make their opossum hold a branch or twig with its tail?

4. Let the children play with their opossums. They can have the opossums climb up and down in the bushes.

Young Opossums Meet a Dog

1. Gather the children on the ground near a bush. Tell a story about the adventures of a hungry, young opossum. Use your opossum, the toy animals (dog, frog or cricket, worm, and nut) and the bush as you act out the story.

   - Very early one morning a tired and very hungry young opossum walks slowly back to its home in the woods when suddenly it sees a frog (or cricket).

   - The opossum opens its mouth to eat the frog, but the frog hops away.

   - The opossum then sees a worm on the wet ground. What do you think the opossum does?

   - The opossum opens its mouth to eat the worm, but the worm slithers down into a little hole. (Or, have the opossum eat the worm if a child suggests it.)

   - The young opossum sees a dog. The dog barks and runs toward the opossum.

   - The opossum quickly climbs up into the bush. The dog tries to reach the opossum but can't, and so the dog slowly walks away.

   - The opossum looks around, climbs down out of the bush, and walks around on the ground. It sniffs for food.

   - It finds a nut. What do you think the opossum does?

   - The opossum eats the nut.

   - The dog sees the opossum again and runs toward it.

   - The opossum does a very funny thing. It rolls over on its back and doesn't move.

   - The dog sniffs the opossum and walks away.

2. Tell the children that sometimes a real opossum lies on its back and stays very still when a dog comes too close. The opossum pretends it is dead, and the dog leaves it alone.

This is the origin of the expression "playing possum."

3. Encourage the children to share their ideas about why the dog leaves the opossum alone when the opossum does its "trick."

4. Encourage the youngsters to pretend they are opossums. Make the toy dog sniff them. They can roll over on their backs and stay very still when the toy dog comes along.

Creative Play

1. Allow time for the children to play freely outside near the bushes with their young opossums and the toy dog. Children can collect a small branch or twig for their opossum to climb on. Have extra toy opossum foods available for the children to use in their dramas.

2. Encourage the children to make up games and dramas about opossum behavior and to perform them for others.

Optional

The Cardboard Tree

Place the cardboard tree in the dramatic play area for the children to use when they play with their opossums; the youngsters can make their opossums climb the tree or live in the tree holes.

Indoor Tree Branch

Bring in a number of small branches, or one large tree branch, to place on the floor or a table for the opossums to climb on as a natural display.

Session 3: Measuring Young Opossums

What You Need

For the group

❑ 1 paper young opossum made during Session 1

❑ Common objects from the classroom that are longer, shorter, and the same length as the paper opossum, such as straws, pencils, markers, books, paintbrushes, stuffed animals, and spoons

❑ White paper for duplication

For each child

❑ 1 paper young opossum made in Session 1

❑ Measurement Hunt at Home letter for parents (see page 57, and there is a tear-out version of the letter at the end of the guide)

Optional

❑ Human baby doll, approximately life-size, 12"–21" long

Getting Ready

1. Place an assortment of objects that vary in length on tables or in an areas that are readily available to the children.

2. Select three different objects from the assortment to demonstrate how to measure using the paper opossum. Choose one object that is the same length, one that is shorter, and one that is longer than the opossum.

3. Duplicate for each child a copy of the "Measurement Hunt at Home" letter describing the measurement hunt to be done at home. See page 57.

Comparing Young Opossums to Familiar Items

1. Gather the children in a circle. Lay the paper opossum flat on the floor. Hold up an item, such as a marker, and compare it to the opossum. Place the end of the marker at one end of the opossum.

2. Ask the children to compare the length of the opossum to the marker.

   Is it shorter than the marker?

   Is it longer than the marker?

   Is it the same length as the marker?

3. Take another item, such as a paintbrush. Put the end of the paintbrush at the end of the opossum. Have the children compare the length of the opossum to the paintbrush.

4. Continue with the third item, such as a book. Hold it up and ask the children to guess if the opossum is longer, shorter, or the same length as the book.

*Be sure to use the words "**shorter than**," "**longer than**," and "**the same length as**" rather than "smaller than," "bigger than," or "same size as" so that length is not confused with size.*

5. Ask them how to compare the length of the opossum to the book. Follow their directions and compare the lengths.

    Which is longer?

    Which is shorter?

More Comparing Young Opossums to Familiar Items

1. Give the children their opossums. Ask them to touch the tip of their opossum's nose. Remind them the nose is the starting point for measuring.

2. Invite the children to the tables or designated areas to use their opossums to measure some of the objects on the table.

3 After about five minutes of measuring and comparing, ask each child to select one item from the table.

4. Have the children bring their items back to the circle area.

5. Let the children take turns showing which items they brought to the circle. Have them measure their items using their opossums. Be sure to remind them they need to line up the object with one end of the opossum.

6. With each child, ask if the opossum is longer, shorter, or the same length as the object they selected. As the children share their measurements, they can sort the items into three groups: "longer than," "shorter than," and "the same length as."

7. As interest permits, invite the children to return to the tables and continue to measure their opossums against other objects on the table.

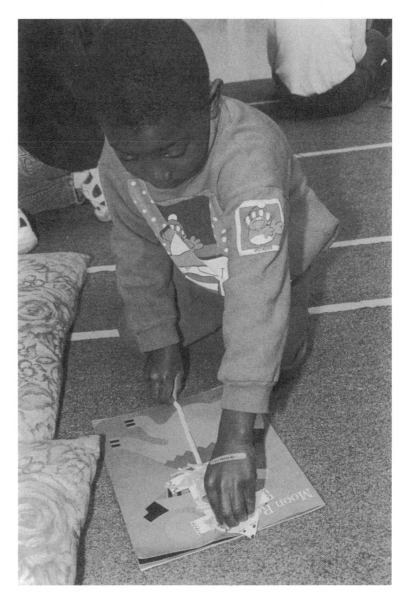

The Measurement Hunt (Home/School Connection)
Directions to Students for Measuring at Home
1. Tell the children they are going on a "Measurement Hunt at Home"—they will look for one or more objects in their homes that are the **same length** as their young opossums.

2. Review how to use the young opossum to measure three objects—one the same length, one longer, and one shorter than the opossum. Review how to use the tip of the opossum's nose as a starting point for measuring.

3. Give the children their young opossums and the letter to their families describing the hunt. Tell them to bring back to class one or more objects from their homes that are the same length as their opossums.

Sharing Objects in the Classroom
1. Gather children in a circle with their objects from home. Have on hand a few extra items for those children who forgot to bring an object from home.

2. Look at the assortment of objects. Give the children an opportunity to report on their measurement hunts.

   Was it easy to find objects the same length?

   What other things did they measure using the opossum?

   Did two children bring in the same object?

3. Provide time for the children to play with their opossums and to use them to continue measurement activities.

Going Further
Comparing Opossums and Human Babies
Bring in a human baby doll that is life-size. Have the children look at the doll and compare it with the newborn opossum and the young opossum. Have them make observations about the babies. Ask questions that guide them to make observations, such as:

What are some characteristics that all share? *[Eyes, hair, nose, mouth]*

What are some differences among them? *[Fur, size]*

Which is longer, the human baby or the young opossum?

Do parents of human babies and parents of opossum babies do any similar things to take care of their children?

Charting Data
If the children's young opossums are the same length, create a chart with three columns listing items that are longer than, shorter than, and the same length as the young opossum.

Measurement Hunt at Home

Dear Families,

We are having a wonderful time studying about opossums. Hopefully, you are hearing reports from your child about our activities. Here is a sketch of what we have learned so far: opossums are nocturnal animals with a keen sense of smell; mother opossums have a pouch to carry and nurse their young; the babies are about the size of a small bean at birth; and the mother can care for up to 13 babies in her pouch. In addition, they have measured a life-size mother opossum with objects, done counting activities with "baby" opossums; and even counted the pockets on their clothing—the closest things to pouches on humans!

Recently, we learned about young opossums and their behavior after they leave the safety of their mother's pouch. Your child made a young opossum out of paper and, today in class, this paper opossum was used as a measurement tool to compare its size to that of common objects in the classroom. We organized the objects the children measured into those "longer than," "shorter than," and the "same length" as the young opossum.

Your child has been asked to go on a "measurement hunt" in your home. They are to find one or more objects that are the same length as their paper opossums. We would like your child to bring back to class one or more objects (but no more than five!) that are the same length as the paper opossum. These items will be returned. However, please do not send to school any breakable or valuable objects!

Though this may seem like a simple activity, it reinforces the measurement concepts in the national mathematics standards for young children. As the children use their paper opossums to measure, they are comparing objects in a familiar setting and using a nonstandard tool to measure the relative length of objects. This work lays the foundation for using (in later school years) standard units of measurement to determine length, area, capacity, weight, and time.

Please assist your child as needed with this activity. Be sure to have them explain to you what they are doing and why.

Thanks for your participation.

Measuring a Young Opossum with Small Objects

Recommended for Kindergarten and First Grade

1. Have the children measure their young opossums just as they measured the mother opossum (Activity 1, Session 4). This time have them use smaller objects to measure, since the young opossum is smaller. For example, paper clips, unifix cubes, pennies, bottle caps, and lima beans, can be used—just be sure all objects of the same type are the same length.

2. If you want to review how to measure, choose an object the children will use. Hold it up and have the children estimate how many of that object it will take to go from the tip of the nose to the end of the tail. Place the objects down one at a time until you are halfway across the length of the opossum. Give them an opportunity to revise their estimates.

    Continue to place objects across the opossum until you reach the end of the tail. Count the objects.

3. Show the children the objects they can choose to measure. Before they measure and count, encourage them to estimate how many of each object it will take to measure the length.

4. Remind them to select only one container of objects at a time to measure the length. When they have completed measuring with that object, they should return it and select another.

5. This can be done as a whole class or at a learning center. As appropriate, have children record the results of their measurements. Since students are revisiting this activity in a new format, it is an opportunity to assess their measurement skills.

Session 4: A Class Book

What You Need

For the whole group

- ❏ 1 Mother Opossum poster
- ❏ 1 paper newborn opossum made in Activity 2: Mother Opossum's Pouch
- ❏ 1 Young Opossum poster
- ❏ Paper, drawing and writing materials for each student

Reviewing from Newborns to Young Opossums

1. Have the children sit in a circle on the floor. Show them the drawing of a Newborn Opossum and the Mother Opossum poster. Ask,

   "How is the newborn opossum different from the mother opossum?"
   [It's smaller; its eyes are closed; it doesn't have fur; it's pink; the mother is white]

2. Ask,

   "Where do the babies stay when they are small?"
   [In their mother's pouch]
   "What do baby opossums do in the pouch?"
   [Drink milk; stay warm; grow bigger]

3. Show the Young Opossum poster to the group. Point to the poster and ask,

   "How is this young opossum different from the newborn opossum?"
   [It's bigger; it has fur; its eyes are open]

4. Ask,

   "What can young opossums do after they leave their mother's pouch?"
   [Ride on her back; climb trees; hold on with their tails; pretend to be dead when something dangerous comes by]

Creating a Class Book

1. Have the children think about what they have learned. Have them draw a picture of something from the unit.

2. For preschoolers, after they finish the picture, have them dictate what they learned and you, or an assistant, can write it on their pictures.

   For kindergartners and first graders, have the students write a sentence or a story about opossums. Assist them, as necessary, with the writing.

3. Put all the pages together in book form and read it to the class.

This is a great time to have upper grade "buddies" assist if your kindergartners have them.

This is an opossum playing dead.

Michael

A Perfect Pocket for an Opossum!

written and illustrated by:

KELCEY   ALEX
BRIAN   WILEY   Kolton
Mi
ALEX   leah
JacK
SARAH

He comes out at night and looks for food

Alex

Opossums hang their babies under their bodies in a pouch.

Brian

Going Further

1. Let the children take their pouches, opossums, and any opossum journals home. Encourage the youngsters to show their friends and family some of the things opossums can do.

2. Share the Class Book with other classrooms or place the book in the school library.

3. Have your students sing the "The Opossum" song to the tune of "She'll Be Comin' Round the Mountain." These new words were written by PEACHES Associate Mary Ehler, director of the Cornerstone Montessori school in Discovery Bay, California.

The Opossum

The opossums been around for a long, long time.
Oh, the opossums been around for a long, long time.
They will eat anything they find.
They have lots of babies, but they don't mind.
Oh, the opossums been around for a long, long time.

The opossums saw the dinosaurs, yes they did.
Oh, the opossums saw the dinosaurs, yes they did.
They scavenge and they hide,
Giving their little ones a ride.
Oh, the opossums saw the dinosaurs, yes they did.

The opossum has five fingers like you and me.
They use them to eat and climb a tree.
Where do the newborns hide?
They find mom's pouch and climb inside.
Oh, the opossum has five fingers like you and me.

The opossum is as friendly as can be.
When's he's little he likes to hang out in a tree.
They won't fight, they use their head.
They'll roll over and play dead.
Oh, the opossum is as friendly as can be.

The opossum is busy all night and sleeps all day.
While you're sleeping he likes to eat and play.
What does he like to eat?
Almost anything's a treat!
The opossum is busy all night and sleeps all day.

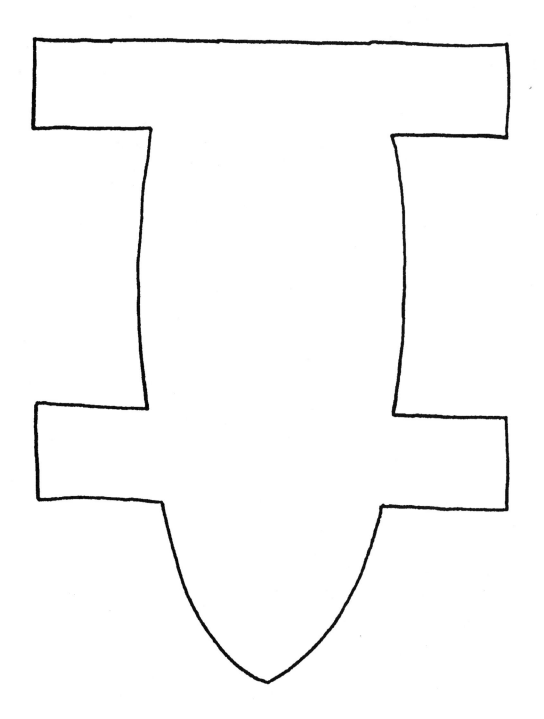

Young Opossum Pattern

©1999 by The Regents of the University of California
LHS GEMS—*Mother Opossum and Her Babies*

Young Opossum

Background Information

It was Captain John Smith (that most fortunate friend of Pocahontas) who named the opossum. The name is from the Algonquian Indian name for the animal—*apasum*—meaning "white animal" or "white beast." Smith described the animal from Virginia this way:

"An Opassum hath an head like a Swine, and a taile like a Rat, and is of the bigness of a Cat. Under her belly she hath a bagge, wherin shee lodgeth, carrieth, and sucketh her young."

Smith's description is not the first, but it is the most famous and most quoted. An early Spanish explorer, Vincent Yáñez Pinzo, picked up a mother opossum and her young after landing on the coast of Brazil around 1500. He took the animal back to Spain and presented it to the monarchs Ferdinand and Isabella. Prior to that, the existence of pouch-bearing mammals was unknown in Europe.

The Virginia Opossum

The opossum that inhabits most of North America, Central America, and part of South America is the Virginia Opossum (*Didelphis virginiana*). It is about the length and weight of a domestic cat. They can be 24- to 32-inches-long, including the 12-inch tail, and can weigh 4- to 12-pounds. Opossums continue to grow throughout their lifetime, so size is a rough determination of age. Females and males grow at the same rate until the stress of nurturing the young slows the females down. As a result, female opossums are slightly smaller than males.

Opossums have black eyes; an elongated snout; pink feet, tail, and nose; and white fur with black tips, which gives them a grayish appearance. They're solitary animals that sleep during the day in a hollow tree or log, a woodchuck burrow, a rocky crevice, a drain pipe, or in a woodpile. They come out at night to search for food.

Opossums don't hibernate, but may remain inactive—wrapped up in leaves, asleep in a den—for short periods during extreme winter weather. Their energy reserves are not extensive, so they must forage on a regular basis throughout the winter.

Where to Find an Opossum

The opossum's distribution across North America is limited northwards by their inability to cope with winter conditions, and westwards by dry, hot climates.

Their distribution, however, is greater now than before Europeans first arrived in the Americas. They now live in about 40 states, as well as parts of Canada. They do well in a large variety of habitats—from sea level up to 9,000 feet, from untouched forests to urban landscapes. Some of their expansion has been due to human activities—their diet can easily include human garbage as well as cultivated fruits and vegetables.

Hereby Hangs a "Tail"

Contrary to the prevailing myth, opossums don't hang upside down by their tails while sleeping, nor do the young hang from their mother's tail when they're old enough to leave the pouch. The opossum's tail is prehensile (adapted for seizing or grasping, especially by wrapping around), but can't support the weight of an adult. Opossums do use their tail as a climbing aid—for balance and support while climbing among tree branches. One will sometimes wrap its tail onto a higher branch while climbing down to a lower branch, giving the false impression it is hanging. The "nakedness" of the tail may be an adaptation for its use in climbing. The rough tail skin makes it very useful for gripping branches.

Opossums can also use their prehensile tail to carry dry grasses and leaves to line their den. They pick up several mouthfuls of the material, push each under the abdomen to the waiting tail, which is curved in a loop, then drag the load to the den.

Life Span

Opossums can live up to 10 years in captivity, but rarely live more than one or two years in the wild, and the females rarely have more than two litters in their short lifetime.

Opossum Feet

On the front feet, opossums have five fingerlike toes that are well-suited for grasping. This makes it possible for opossums to manipulate their food, grab a tree branch, or for the young to hold on to their mother's fur while riding on her back.

On the back feet they have an opposable toe that resembles a thumb, making grasping branches an easy task—and opossums great climbers. The only other type of mammal with a thumb are primates. The opossum's "thumb" can meet any of the other four toes of the same foot, acting therefore as a very flexible "super thumb." The four toes on the back feet have sharp claws, but the "thumb" has no nail—perhaps a further adaptation for grasping. While opossums don't hang from branches by their tails, they can hang by their back feet.

That Handy Pouch

The Virginia Opossum is North America's only marsupial. Like the marsupials of Australia—kangaroos, koalas, wombats, and wallabies—female opossums have a pouch. In this fur-lined pouch, the young opossums find food, warmth, and protection. The pouch, like a huge pocket, provides an ideal way for the females to carry their young from place to place as they search for food or shelter. The pouch is held shut by muscles, keeping the young safely inside.

Opossum Young

After a mere 13-day gestation period—the shortest of any mammal—the female opossum gives birth to as many as 20 newborns. Just before her babies are born, the female sits up and licks the area between the birth canal and the pouch to provide "a path" for the young to follow. These embryonic young, each only half-an-inch long (about the size of a raisin or bean), use their well-developed front legs and claws to climb into the pouch. During this two-inch trip, the young receive no help from their mother, except that she sits hunched over to make the distance as short as possible. Only the strongest survive the trip to the pouch, and since mother opossums have only 13 nipples, only the first 13 to arrive find a nipple. Each baby opossum stays attached to a nipple for 60 days.

Newborn opossums are pink and hairless, and their eyes are closed. By the time they are 45 to 60 days old, the young opossums have fur, whiskers, and their eyes are open. They begin to move outside the pouch for brief periods but return to nurse. After about three months, they are weaned. The young stay with their mother—joining her on nightly foraging trips—until they are three or four months old. After that, they're on their own. For up to three months after

Number and

Measurement begins with comparing. And, as a way to communicate comparisons, we use units of measurement. For young children, it is important to begin with nonstandard units of measurement, such as blocks, paper clips, and other familiar objects. Using these objects as their measurement tool, children gain experience with the concept of measurement and units of measurement. This foundation enables them to later make the transition to standard units of measurement.

Number

For many people, mathematics is synonymous with numbers and arithmetic. Numbers do permeate all of mathematics. For starters, numbers are used to define quantities and relationships, to measure, to make comparisons, to interpret information, and to solve problems. Beginning with concrete, real-world experiences, numbers are understood through counting "how many" and/or measuring "how much."

In this guide, children develop an understanding of numbers through concrete experiences. As they predict how many objects it will take to cover the length of the Mother Opossum poster, they develop estimation skills. Through this estimation process and counting the number of objects used to cover the length of the Mother Opossum poster, number sense is developed.

Numbers are written in the base 10 system using the numerals 0 through 9. As children record numbers in the context of counting, they make the connection between the concrete quantity and the abstract representation of that number. In order to comprehend the base 10 system, children first need to understand the order of numbers, and how numbers relate to one another. Building on that, they need concrete experiences grouping objects into "tens" and "ones." In this unit, as the children count pockets using unifix cubes, they group the cubes by "tens" and "ones," and record the number representing the leaving their mother, litter mates may share common dens. At six to eight months the young are ready to start their own families, but usually don't until a year after they're born.

Opossum Food

Opossums do not have any way to store food or energy and need food sources that are stable throughout the seasons and from year to year. Luckily, opossums eat almost everything, including insects, earthworms, snails, spiders, salamanders, frogs, crayfish, lizards, snakes, and mice. They also eat eggs, birds, young rabbits, and squirrels. They raid garbage cans and chicken coops, and dig up vegetable gardens. They eat grasses, plants, seeds, nuts, corn, and fruits, especially persimmons, apples, and grapes. Opossums are not territorial, so when food resources become depleted in one area, they simply move to another area. They are sometimes seen on roads at night eating animals killed by automobiles.

Measurement

grouping. These concrete experiences lay the foundation for conceptual understanding of numbers and the base 10 system.

Measurement

Measurement is a process we use constantly: in cooking we measure both wet and dry items to prepare recipes; at gas stations we fill our tanks by the gallon, oil is put in our cars by quarts; and to organize our daily schedule, clocks and calendars let us know the date and time. Length, capacity, weight, mass, area, volume, time, temperature, angle, and money all make use of measurement.

Measurement begins with comparisons—which item is heavier? smaller? longer? wider?—something children like to do! As children make comparisons, they investigate the attributes of length, weight, mass, volume, area, capacity, and time. To investigate a specific measurement, tools are needed. At first, nonstandard, familiar items, best serve as the measurement tools—toothpicks, spoons,

wooden cubes—to develop the process of measuring using a consistent unit. As the children measure the Mother Opossum poster, they make predictions about how many of each unit it will take to cover the length of the opossum. This connects measurement with number concretely and helps develop their number sense.

These experiences lay the foundation for using standard units of measurement and systems of measurement. When students encounter the metric system, English system, currency system, and so on, they will draw on their experiences of the process of measurement and the use of a unit of measure and apply it to other standard systems. Furthermore, they will see that all of these systems allow us to speak a universal language—when someone asks for a yard (or meter) of fabric, the quantity is defined and understood. Eventually, students will be able to choose the most appropriate unit and measurement tool for each particular situation they encounter.

Enemies

As their forest habitat decreases, opossums are forced into closer association with people, who are the biggest threat to their survival. Often while eating road kill, opossums themselves are killed by automobiles. Opossums sometimes die from insecticides. Some people trap and hunt opossums for their fur, their meat, and because they are often a nuisance to farmers.

In spite of these dangers, people have unwittingly helped opossums by drastically reducing the numbers of two of their major predators, coyotes and wolves. Other animals that kill opossums are dogs, foxes, bobcats, owls, hawks, snakes, and snapping turtles.

Defenses

Opossums are usually passive, gentle animals. When threatened, they defend themselves in numerous ways. They may bare their 50 needle-sharp teeth, hiss, and growl, or they may scramble away and climb a tree. The opossum's best known defense is pretending to be dead, commonly known as "playing possum." When confronted by a predator, an opossum may collapse on its side, open its mouth in a strange grimace, drool, exude a distasteful smell from its anal glands, and remain motionless until the predator, uninterested in such "lifeless" and smelly prey, moves away. This is an effective defense since many predators eat only the food they kill. The opossum may remain motionless for just a few minutes or several hours. Many people think opossums faint from fright, but recent experiments with an encephalogram, a machine that records brain activity, show that these motionless opossums are in a highly alert mental state.

ONCE UPON A TIME THERE WAS AN OPOSSUM AND SHE HAD ONE BABY AND SHE WAS FEELING SAD AND ANOTHER BABY CAME AND SHE WAS SO HAPPY AND IT WAS SNOW SO SHE TRAVELED TO MEXICO BECAUSE IT WAS SO MUCH WARMER THERE SO SHE WANTED TO STAY THERE FOREVER

THE END

CAITLYN

Assessment Suggestions

Selected Student Outcomes

1. Students improve their understanding of size as they compare adult and young opossums to their own body and body structures.

2. Students develop measurement skills as they use nonstandard units to measure length.

3. Students develop their number sense by estimating and counting in context.

4. Students become familiar with different opossum behaviors related to feeding, defense, and parenting.

5. Students are able to describe opossum body structures, such as the pouch, paws, tail, and their functions for opossum survival.

6. Students can describe stages of opossum growth and development from birth to adulthood and compare them to human development.

Built-in Assessment Activities

What Size Are You?

In all three main activities, students are introduced to the concept of size as they compare their own size to an adult, newborn, and young opossum; and their hands and feet to opossum paws in activities that use posters and models. The teacher observes how students participate in the activities and listens to their responses to questions regarding size. (Outcome 1)

Using Young Opossums to Measure

Students use their models of young opossums as a nonstandard tool to compare the length of the opossums with familiar objects. The teacher observes them as they measure to see how well they use the process of measurement to compare the relative lengths of objects. (Outcomes 1, 2)

Measuring Mother Opossum

In Activity 1: Getting to Know Opossums, students measure the length of a drawing of a life-size mother opossum using familiar objects found in the classroom. Later in the unit, they also measure the length of their young opossum models. In using nonstandard units to measure, and later recording results on a data sheet, students practice important measuring skills, as well as estimating and counting. The teacher observes how well students perform the activities and notes their accuracy in measuring and recording. (Outcomes 2, 3)

Counting Pockets and Baby Opossums

In "How Many Babies in Mother Opossum's Pouch?," students glue 13 beans (representing newborn opossums) onto a drawing of a mother opossum with 13 nipples. This reinforces one-to-one correspondence and counting to 13. They play a game in which they estimate and count the number of babies in their simulated pouches. In "Counting Our Pockets," students estimate the total number of pockets on everyone's clothing, and then count the number a variety of ways, including by tens and ones. In each activity, the children's estimates and their counting skills provide an insight into their number sense. (Outcome 3)

What Opossums Do

Throughout the unit, students participate in dramatic play and role-play as they act out opossum behaviors, including sniffing for food, "playing possum," and having baby opossums ride in the mother's pouch or on her back. During these activities, the teacher listens for descriptive language, explanations, and questions as students use play to communicate important opossum behaviors. (Outcomes 4, 5, 6)

Babies Grow Up

In Activity 2: Mother Opossum's Pouch and Activity 3: Young Opossums, students create paper models of a pouch and a young opossum that they use in dramatic play. The teacher observes the processes the children use in making the models, the finished models, and play activities, taking note of the ideas and questions generated. Responses to specific questions about opossum growth and human growth are also noted. (Outcomes 5, 6)

Additional Assessment Ideas

Opossum Stories

Have students write and dictate stories or create an Opossum Journal. (Outcomes 1, 4, 5, 6)

Measurement Fun

Collect other life-size posters of animals and have students measure them with nonstandard units or compare the animal body structures to their own. (Outcomes 1, 2, 5)

Button Count

Have students estimate the total number of buttons on their clothing and count using a tool for measurement, such as unifix cubes. (Outcome 3)

Dramatic Play

Create a "woodland" play area with props where children can pretend to be opossums or other forest animals caring for their young. (Outcomes 4, 5, 6)

Animal Parents

Read stories about other animals that have different ways of caring for their young. Encourage children to discuss the similarities and differences of these animals to both opossum and human parenting behaviors. (Outcomes 4, 5, 6)

Resources

Through your local wildlife rehabilitation center, nature center, or science museum, you may be able to arrange for a classroom visit by a docent or other staff person to talk about opossums and bring a live opossum or a mounted specimen to your classroom. Your county or state fish and game agency may also be able to help.

— Books —

Kangaroos and Other Marsupials
by Norman Barrett
Franklin Watts, New York, 1991

Although the book contains no information specifically on the Virginia Opossum, it does have good color pictures showing the variety of other marsupials—such as kangaroos, opossums, and wombats—with descriptions of their physical characteristics, habits, natural environment, and evolution.

Kangaroos, Opossums, and Other Marsupials
by Marie M. Jenkins, illustrated by Matthew Kalmenoff
Holiday House, New York, 1975

The history and characteristics of marsupials are examined with specific examination of the habits and characteristics of opossums, kangaroos, and other similar animals.

Mathematics Their Way
by Mary Baratta-Lorton
Addison-Wesley Publishing, Menlo Park, Calif., 1976

This invaluable resource for teachers of primary students includes activities and the rationale for teaching patterns, number sense, estimation, counting, computation, place value, logic, sorting and classifying, comparing, measurement, and graphing.

Meet the Opossum
by Leonard Lee Rue III with William Owen
Dodd, Mead, New York, 1983

The physical characteristics, behavior, habitat, food habits, and babies of the opossum, which has inhabited the Earth since the time of the dinosaurs, are observed using black and white photos.

Nature's Children: Opossum
by Laima Dingwall
Grolier, New York, 1986

This is a fairly thorough natural history of opossums with good color photos, including several of young opossums in the pouch.

The Opossum

by Emily Crofford

Crestwood House, New York, 1990

> With good color photographs of the opossum young in the pouch and at different ages, this book examines the physical characteristics, behavior, and natural environment of the opossum.

Opossum

by Kazue Mizumura

Thomas Y. Crowell, New York, 1974

> In a story-like fashion using illustrations, this book describes the characteristics of the opossum and its ability to play dead to evade predators.

The Opossums

by Anne LaBastille

The National Wildlife Federation, Washington, D.C., 1973

> In two stories, Ranger Rick and his friends point out the danger of phosphate detergents and look for missing opossum babies. The story is full of factual information and color pictures of opossums.

The World of the Opossum

by James F. Keefe, photos by Don Wooldridge

J.B. Lippincott Company, Philadelphia, 1967

> This thorough discussion of the life and development of the opossum includes a very informative description of the birth of the young and how they travel to the pouch. There are many good black and white photos.

— Magazines —

Three of the magazines published by the National Wildlife Federation—*Your Big Backyard, Ranger Rick,* and *National Wildlife*—contain many articles about opossums. Your school or local library may have several years worth of back issues.

— Music —

"Keepers of the Animals"

Fulcrum Publishing, Golden, Colorado, 1992

> This two-cassette set is a collection of 24 Native American legends demonstrating the power of animals in Native American traditions. The stories are told by Joseph Bruchac and include the delightful "Why Possum Has a Naked Tail." The book, *Native American Animal Stories*, which contains these stories, can be found in the Literature Connections section on page 74.

"Animal Folk Songs for Children"

Rounder Records, Cambridge, Massachusetts, 1992

> This collection contains more than 50 American folk songs about all kinds of animals, including the opossum. Mike, Peggy, Barbara, and Penny Seeger and their children perform in tradition-based styles on banjo, guitar, piano, fiddle, tin whistle, ukulele, mandolin, jaw harp, lap dulcimer, quills, and a variety of other acoustic instruments. It is available on cassette or CD.

"Wake, Snake: Children's Stories and Songs of the South"

J.J. Reneaux, August House, Little Rock, Arkansas, 1996

In this pleasant mix of stories and songs, Reneaux gives the listener a happy taste of the South. She sings the songs and tells the stories well and even throws in several personal experiences. All flow together smoothly. The tape includes the Cherokee story of why possum has a bare tail as well as a possum song written by Reneaux.

— Puppets/Stuffed Animals —

Opossum

13" long stuffed opossum and a 7" long baby opossum finger puppet (item #T2134)

Folktails by Folkmanis
1219 Park Ave.
Emeryville, CA 94608
(510) 658-7678

Acorn Naturalists also carries a 13" opossum with a 7" velcro-attached baby. It is item number P-2536 and Acorn can be reached via acornaturalists.com or at (800) 452-2802.

— Opossum Societies —

Opossum Society of the United States
P.O. Box 16724
Irvine, CA 92713

The National Opossum Society
P.O. Box 3091
Orange, CA 92857-0091

— Internet —

The Wonderful Skunk and Opossum Page

http://granicus.if.org/~firmiss/m-d/md-main

A lot of good opossum information is available here—everything from a brief history of the discovery of opossums to myths, fables, legends, and modern-day books about opossums.

The National Opossum Society

http://www.teleport.com/~opossums/

There is an opossum quiz and a page of facts—both of which are very informative—as well as the principles, goals, and benefits of membership in the society.

Interesting Stuff About Opossums

http://bmewww.eng.uab.edu/BME/MORE/personalities/students/schroeder/opossum.htm

This site contains just what it says it does, plus some great pictures. It is by a member of the Opossum Society of the United States.

— Materials —

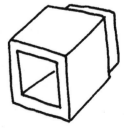

Unifix Cubes

These cubes can be ordered through various teacher supply catalogs.

Math Learning Center	Teaching Resource Center	Didax	Lakeshore
1-800-575-8130	1-800-833-3389	1-800-458-0024	1-800-421-5354
Fax: 1-503-370-7961	Fax: 1-800-972-7722	Fax: 1-800-350-2345	Fax: 1-310-537-5403
Website: www.mic.pdx.edu	Email: trc@trcabc.com		

Literature Connections

Animal Tracks

by Arthur Dorros
Scholastic, New York. 1991
Grades: K–3

This book makes a game out of guessing whose tracks or signs are shown along a sandy shoreline. After the tracks are shown, the reader turns the page to see the animal responsible for them, and then is asked to guess about another set of tracks. This book is a great way to extend the concept of how tracks and signs can tell about the lives and identity of animals.

Animals of the Night

by Merry Banks;
illustrated by Ronald Himler
Charles Scribner's Sons, New York. 1990
Grades: P–1

Simple text and warm rich watercolor illustrations portray the activities of animals that are active at night. Several animals are shown coming out of their daytime resting places to go about their nightly rituals. This is a good introduction to nocturnal animals and shows students some of the things that occur while they're in bed at night.

Don't Laugh, Joe!

by Keiko Kasza
G.P. Putnam's Sons, New York. 1997
Grades: P–1

One can't help but laugh along with this story of a mother opossum who is struggling to teach her giggly son Joe an important lesson—how to play dead. A grumpy old bear helps him learn the lesson, but then Joe is surprised by the bear's intentions. The illustrations enhance this well-told tale.

How Chipmunk Got Tiny Feet:
Native American Animal Origin Stories

collected and retold by Gerald Hausman;
illustrated by Ashley Wolff
HarperCollins, New York. 1995
Grades: K–4

This nicely illustrated collection of stories contains "How Possum Lost His Tail," which explains certain characteristics of the opossum and other animals. The other six origin tales in the book are about how various animals came to be what they are today.

Inch by Inch

by Leo Lionni
Astor-Honor, New York. 1960
Grades: P–1

An inchworm measures a variety of birds and cleverly escapes the ones that want to eat him. A delightful, peaceful story that's useful for demonstrating alternative ways to measure objects.

Katy No-Pocket

by Emmy Payne; illustrated by H.A. Rey
Houghton Mifflin, Boston. 1944
Grades: P–1

This is the classic story of Katy, a mother kangaroo who has no pocket in which to carry her son. To get some ideas on what to do, she asks many other animals how they carry their young. The wise owl advises her to go to the city where Katy finds the perfect pocket.

Native American Animal Stories

told by Joseph Bruchac
from *Keepers of the Animals*
by Michael J. Caduto and Joseph Bruchac
Fulcrum Publishing, Golden, Colo. 1992
Grades: All Ages

There are stories about creation and celebration, and learning lessons. In "Why Possum Has a Naked Tail," one learns not just about the tail, but also about why opossum plays dead, has a silly grin, and the perils of being boastful. See the Resources section on page 72 for a set of audio cassettes with all the stories.

Opossum and the Great Firemaker: A Mexican Legend
by Jan M. Mike;
illustrated by Charles Reasoner
Grades: 1–5

Recounts the Cora Indian story in which Opossum outwits the larger and more powerful Iguana and returns the stolen fire to the people of the Earth.

Opossum at Sycamore Road
by Sally M. Walker;
illustrated by Joel Snyder
Soundprints, Norwalk, Connecticut. 1997
Grades: K–3

Part of the Smithsonian's Backyard series, this one is the adventures of an opossum mother and her young as they travel through the backyard of a house. As the family searches for food they tangle with trash cans and a big brown dog. The book can be purchased alone or with a stuffed, toy 12-inch-mother and six-inch-baby opossum.

A Pocket for Corduroy
by Don Freeman
Viking Press, New York. 1978
Grades: P–2

While at the Laundromat with his good friend Lisa, Corduroy decides he wants to find a pocket for himself. He wanders about and becomes lost, but has a grand adventure. The next morning, when Lisa finds him, she gives him a pocket and another special gift.

Possum and the Peeper
by Anne Hunter
Houghton Mifflin, Boston. 1998
Grades: K–3

When Possum is awakened on the first warm day of spring by a loud noise that won't stop, he and several other animals set out to discover who is making all the racket. In the end, the animals are glad to be awake to enjoy the smells, sounds, and sights of spring. The animals locate the sound using their ears, much like the students identify and find foods using their noses.

possum baby
by Bernice Freschet;
illustrated by Jim Arnosky
G.P. Putnam's Sons, New York. 1978
Grades: P–2

This gentle story chronicles the life of a timid opossum. At first hesitant to leave the warmth and safety of his mother's pouch, the young opossum eventually learns to get along in his new world. As the story unfolds quite a bit of information about opossums is conveyed.

Possum's Harvest Moon
by Anne Hunter
Houghton Mifflin, Boston. 1996
Grades: K–3

When Possum awakes one autumn evening to see a harvest moon, he decides to have a party to celebrate the beautiful moon one last time before the long winter. At first his animal friends decline his invitation, but then realize Possum has a great idea. The animals all have a wonderful time singing and dancing at the party, then retire for the long winter.

Ten Beads Tall
by Pam Adams
Child's Play, Auburn, Maine. 1993
Grades: P–1

This is an excellent book to demonstrate how things can be measured in nonstandard ways. Attached to the book is a string of square beads which are used to measure the illustrations. Statements and questions on each page challenge the reader.

Whose Footprints?
by Molly Coxe
Thomas Y. Crowell, New York. 1990
Grades: P–1

When a mother and daughter go on a walk across their farm through the snow, they discover many sets of tracks and have fun identifying them. It is a pleasant and peaceful story that connects well with the activities about the opossum's feet.

Summary Outlines

Activity 1: Getting to Know Opossums

Session 1: Introducing Opossums

The Opossum Drama

1. Gather the class in a circle, partially darken the room, and pretend to be an opossum (but don't tell the students what you are yet) as you tell them a story of an animal that comes out of the woods at night and looks for food. It finds a cat food bowl, which it pushes around, and eats the food. Ask the children if they know what animal it is.
2. Accept all answers and use the answers to refine your description of an opossum.

The Mother Opossum Poster

1. Show the Mother Opossum poster to the children and ask, "What is this"?
2. Tell them it is an opossum. Ask questions about it. Have the children find its nose, count its legs.

Role Playing Opossums

1. Have the children pretend to be opossums, crawl around the room, find the bowls of food with items that smell you placed in the room earlier, and sniff the contents.
2. Have the children try to identify what was in the bowls and tell you which ones they liked, didn't like, and what was their favorite.

Session 2: Young Opossum Snacks

Young Opossums Sniffing for Food

1. Discuss how opossums have a good sense of smell.
2. Tell the students they are going to guess, by smell, the contents of various containers.
3. Darken the room and present three scent containers. Have the children sniff them and guess what is inside. Give hints if necessary. After the guessing is over, open the containers to see the contents.

Eating Opossum Snacks

Let the children eat, as a snack, the same kinds of food they sniffed in the containers.

Session 3: Mother Opossum and Me *(for preschoolers)*

Comparing Mother Opossum and Me

1. Show the Mother Opossum poster. Have the children find the opossum's nose and end of its tail. Ask the children if they are longer or shorter than the opossum.
2. Place the poster on the floor and have each child measure themselves against it. Line up the top of the child's head to the tip of the opossum's nose. Have the child say whether they are shorter or longer than the opossum.

Session 4: Measuring Mother Opossum with Common Objects

Measuring Mother Opossum Together with Blocks

1. Place the Mother Opossum poster on the floor and gather the children around it. Have them find its nose and end of its tail.
2. Hold up a block and ask how many blocks it will take to cover the opossum from its nose to the end of its tail. Have the children guess.
3. Have the children count with you as you place the blocks on the poster. Line up the first block with the tip of the opossum's nose. At the halfway point, have the children guess again how many blocks are needed to cover the opossum.
4. Finish placing the blocks and count with the children how many blocks there are.

Measuring Mother Opossum Together with Other Objects

1. Choose another object and show it to the children.
2. Remove the blocks from the Mother Opossum poster, but keep one handy. Put an object on the poster at the nose and ask the children if it will take more or less of that object to cover the poster than it took of the blocks. Get estimates from the children on how many objects it will take to cover the poster.
3. Begin covering the opossum drawing with the object (have the children count with you as you put each object down) until you are halfway across the poster. Ask for any revised estimates of how many objects it will take to cover the poster.
4. Finish covering the poster with the object and count them again with the children. See how the actual count compares to the estimates and to the number of blocks needed to cover the poster.

More Measuring

1. Have the children measure the poster for themselves using one type of object at a time. Kindergartners and first graders should work in pairs and record their measurements, if you choose. Have the children make predictions before they measure.
2. Meet together with everyone to discuss results. Compare measurements. If necessary, re-measure the poster with the children.

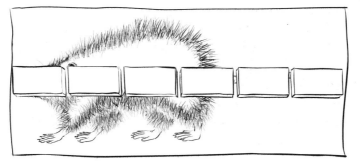

Session 5: An Opossum's Paws

Comparing Front Paws and Hands

1. Show the children the Mother Opossum poster and ask what you call the opossum's feet. Have them count the paws.
2. Give each child a copy of the Opossum Front Paw drawing. Have each child place their right hand next to the opossum paw drawing and compare it to their own hand: is it bigger, smaller, how many toes, how like your hand, how does the opossum use its paws, how do you use your hands and your feet.

Paw Prints Meet Hand Prints

1. Have each child place their hand in a tray of paint and press it on the Opossum Front Paw drawing next to the paw.
2. Ask questions to encourage comparison between hands and paws.

Comparing Back Paws and Feet

1. Give each child a copy of the Opossum Back Paw drawing. Ask them what it is.
2. Have them take off their socks and shoes and compare their feet to the opossum's back paw.
3. Have each child place their foot in a tray of paint and press it on the Opossum Back Paw drawing next to the paw. Ask questions encouraging comparisons between feet and the opossum's back paw.

Activity 2: Mother Opossum's Pouch

Session 1: What's Inside the Pouch?

Pockets

1. Ask questions to encourage talking about pockets: who has a pocket, what do you keep inside it, find all the pockets on your clothes, how many do you have.
2. Ask if they know any animals with pockets and what do they use them for.

Teeny, Tiny, Newborn Opossums

1. Tell the children mother opossum has a pouch much like a pocket. Tie the paper-bag pouch around your waist and tell the children you are pretending to be a mother opossum.
2. Ask, "What do you think mother opossums keep in their pouches?" "How big do you think baby opossums are when they are born?"
3. Give each child a paper newborn opossum to hold. Ask questions to encourage the students to observe the newborns such as its size, eyes shut or closed, its tail and feet.
4. Tell the students the babies are born with their eyes shut and they crawl into their mother's pouch. Have the children crawl their babies one at a time into your pouch, counting the babies along the way.
5. Ask what the babies do in the pouch. Explain about getting milk from the mother opossum's nipples and that the pouch is safe and warm for the babies.

Wearing Opossum Pouches

1. Tie a pouch on each child who wants one, and give them each one or more babies. Let the children play freely with their newborn opossums and pouches.
2. Let them wear the pouches during the day keeping the newborns safe in the pouches. Ask questions to review what the children learned about baby opossums.
3. Save the pouches for later use.

Session 2: How Many Babies in Mother Opossum's Pouch?

How Many Babies Are in the Pouch?

1. Show the children a pink bean. Tell them it is the same size as a baby opossum. Ask where the babies crawl after they are born.
2. Show the children the Mother Opossum's Pouch sheet. Count the nipples with the children. Tell them 13 is the most babies the mother opossum can have in her pouch.
3. Write your name on the pouch sheet and have the children count with you as you glue 13 beans onto the nipples on the sheet. Write the number 13 on the sheet.
4. Have the children glue 13 beans onto their own Mother Opossum's Pouch sheet. Have them write their names and the number 13 on the sheet.

Counting Bean Babies

1. Tie a pouch around your waist and put 13 beans in it. Tell the children to pretend the beans are baby opossums and you are a mother opossum. Ask, how many babies can a mother opossum have in her pouch?
2. Take out a handful of beans and count them with the children. Put them back, take out another handful and count those.
3. Give the children their pouches with 13 beans and have them take a handful out and count them with a partner.

Estimating Bean Babies *(for kindergartners and first graders)*

1. As above, but have the children estimate the number of bean babies before they count them.

Session 3: Counting Our Pockets

Counting Pockets

1. Ask for a way to count everyone's pockets. Tell the students you will use unifix cubes as a counting tool. Model how to do this. Put one cube in each of your pockets.
2. Have the children also put one cube in each of their pockets.
3. Remove your cubes and link them together to make a "train." Have the children do the same with their cubes.
4. Compare the "trains" for length (shortest, longest, same length).
5. Make statements about the trains and have the children match their train to the statement: Zero cubes in your train, raise your hand; three cubes, raise your hand, etc. If you have an odd/even number of cubes, raise your hand; the most number of cubes, raise your hand. Count the cubes in the longest train.
6. Remind the students the cubes represent the number of pockets they have.

7. Take your train and have a child add her train to it; continue until all the trains are joined together. Have the children estimate how many cubes there are all together several times as you join all the trains. Tell the children this is how many pockets everyone has. Once finished, count, by ones, the number of cubes in the "train."

Counting Pockets Other Than by Ones

1. Ask if there are other ways to count than by ones. Count based on suggestions.
2. Count the "train" by 10's. Break off 10 cubes. Line those 10 cubes against the long "train" and beak off another 10. Continue until there are less than 10 cubes left in the "train." Break off the remainder by ones and count the entire "train" again by 10's and ones.
3. If appropriate, count the "train" by fives.
4. Tell the children they will do the activity the next day and have them guess if there will more or less pockets, or the same.

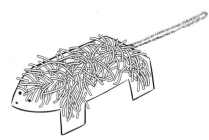

Activity 3: Young Opossums

Session 1: Little Opossums Grow Up

Making Young Opossums

1. Show the children the Mother Opossum poster and encourage them to talk about what the know about opossums.
2. Tie a pouch around your waist and have the children guess what is in your pouch. Climb a newborn opossum out of the pouch. Tell the children the baby drank lots of milk and grew. Climb the paper young opossum, you made earlier, out of the pouch and ask how it is different from the baby.
3. Ask what the opossum needs and as the children respond, draw eyes, pink nose, whiskers, ears, and claws. Turn the opossum over and write your name on it.
4. Glue on yarn and tell the children the opossum is growing fur.
5. From your pouch give each child a paper young opossum. Have them draw features, glue fur, and write their names on it.
6. Show the children the Young Opossum poster and compare it to the Mother Opossum poster and the paper opossum they made.

Session 2: A Young Opossum's Tricks

Riding on Mother Opossum's Back

1. Give each child their pouch with their young opossum in it. Ask them how the mother opossum carries the young opossums when they are too big to ride in the pouch. Crawl the young opossum out of your pouch and onto your back.
2. Ask how it hangs on. Let the children play with the young opossums, crawling them out of their pouches and riding them around on their backs.

Young Opossums in Trees

1. Take the children, with their pouches and young opossums, outside to a bushy area.
2. Climb your opossum out of your pouch and onto a branch. Tell the children opossums like to climb trees. Have the youngsters play with their opossums in the bushes and show them how the opossum hangs on with its tail to a branch.

Young Opossums Meet a Dog

1. Gather the children and tell them a story of a young, hungry opossum. It tries to eat a frog, or cricket, and a worm, but they all escape. A dog comes along and the opossum plays dead until the dog goes away. Tell the children opossums trick other animals this way. Have them take turns pretending to play dead as you pretend to be the curious dog.

Creative Play

Let the children play freely with their young opossums and encourage them to create opossum dramas and games.

Session 3: Measuring Young Opossums

Comparing Young Opossums to Familiar Items

Lay the paper young opossum on the floor and compare its length to an item, such as a marker. Line up an end of the marker to the tip of the opossum's nose. Ask the children comparison questions: which one is longer, shorter, same length. Do the same with several more items.

More Comparing Young Opossums to Familiar Items

Give the children their own young opossums. Have them measure it against familiar items from a central area and then have each child pick one item to show the group. Ask each child whether their item is longer, shorter, or the same length as their paper opossum.

The Measurement Hunt (Home/School Connection)

1. Tell the students they are going on a measurement hunt at home. Review how to measure items against their paper opossums. Send the Measurement Hunt letter home. Tell the students to bring from home one or two items that are the same length as their opossum.
2. Using the items from home, have each student report on what they found: was it hard to find something the same length, did they measure lots of items, did several children bring the same or similar items from home.

Session 4: A Class Book

Reviewing from Newborns to Opossums

Show the children the Newborn Opossum drawing, the Mother Opossum poster, and the Young Opossum drawing. Review where newborn opossums stay, how the babies change as they get older, and opossum behavior.

Creating a Class Book

1. Have each child draw a picture of what they've learned. For preschoolers, have them dictate something they learned. For kindergartners and first graders, have them write something they learned next to their pictures.
2. Assemble all the pages into a class book and read it to the children.

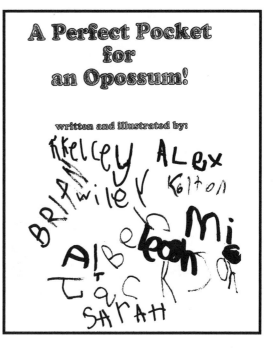

A Perfect Pocket for an Opossum!

written and illustrated by:

Opossum Front Paw

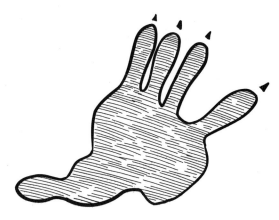

**Opossum
Back
Paw**

©1999 by The Regents of the University of California
LHS GEMS—*Mother Opossum and Her Babies*

RECORDING SHEET

MEASURING TOOL	✓ ACTUAL ✓

RECORDING SHEET

MEASURING TOOL	? PREDICTION ?	✓ ACTUAL ✓

Mother Opossum's Pouch

©1999 by The Regents of the University of California
LHS GEMS—*Mother Opossum and Her Babies*

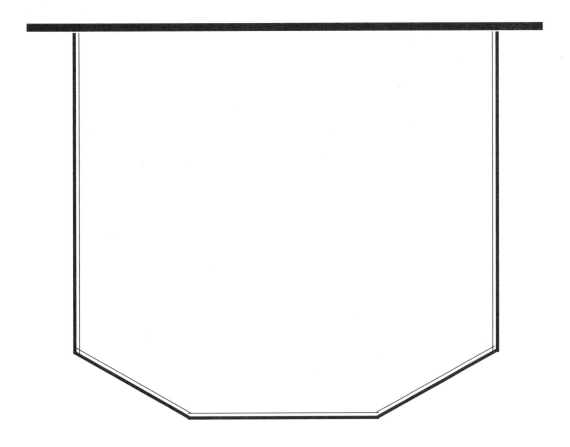

©1999 by The Regents of the University of California
LHS GEMS—*Mother Opossum and Her Babies*

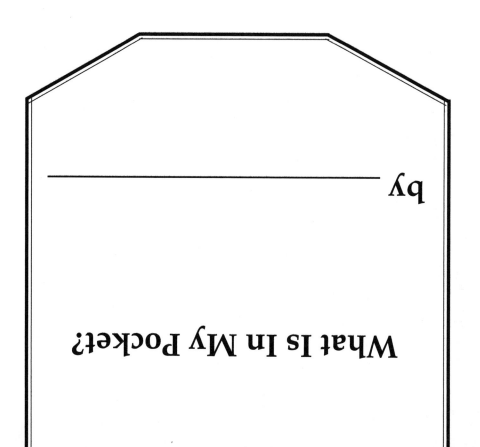

by _____

What Is In My Pocket?

©1999 by The Regents of the University of California
LHS GEMS—*Mother Opossum and Her Babies*

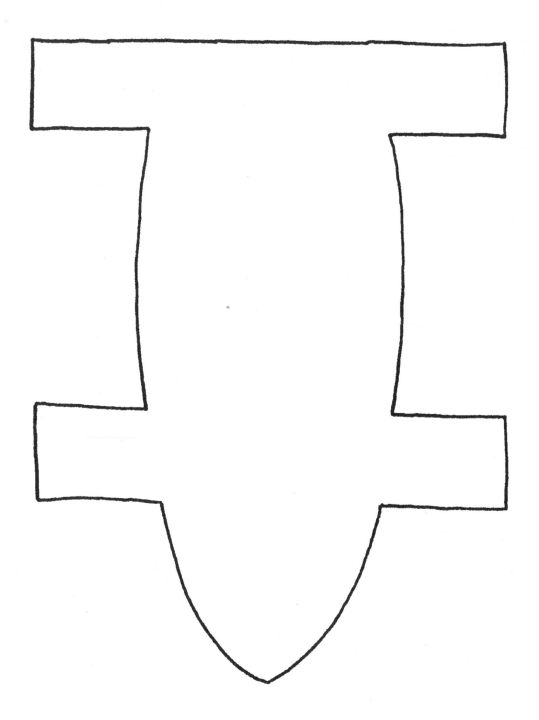

Young Opossum Pattern

Young Opossum

Measurement Hunt at Home

Dear Families,

We are having a wonderful time studying about opossums. Hopefully, you are hearing reports from your child about our activities. Here is a sketch of what we have learned so far: opossums are nocturnal animals with a keen sense of smell; mother opossums have a pouch to carry and nurse their young; the babies are about the size of a small bean at birth; and the mother can care for up to 13 babies in her pouch. In addition, they have measured a life-size mother opossum with objects, done counting activities with "baby" opossums; and even counted the pockets on their clothing—the closest things to pouches on humans!

Recently, we learned about young opossums and their behavior after they leave the safety of their mother's pouch. Your child made a young opossum out of paper and, today in class, this paper opossum was used as a measurement tool to compare its size to that of common objects in the classroom. We organized the objects the children measured into those "longer than," "shorter than," and the "same length" as the young opossum.

Your child has been asked to go on a "measurement hunt" in your home. They are to find one or more objects that are the same length as their paper opossums. We would like your child to bring back to class one or more objects (but no more than five!) that are the same length as the paper opossum. These items will be returned. However, please do not send to school any breakable or valuable objects!

Though this may seem like a simple activity, it reinforces the measurement concepts in the national mathematics standards for young children. As the children use their paper opossums to measure, they are comparing objects in a familiar setting and using a nonstandard tool to measure the relative length of objects. This work lays the foundation for using (in later school years) standard units of measurement to determine length, area, capacity, weight, and time.

Please assist your child as needed with this activity. Be sure to have them explain to you what they are doing and why.

Thanks for your participation.